Maths Workshee

GW01458266

Maths Worksheets is a book of photocopy masters providing inform............. ....
maths for adult students. In the first part *(pp. 4-32)*, the worksheets are headed according
to the specific maths skill covered. In the second part *(pp. 33-44)*, they contain more general
exercises on a particular everyday topic. The worksheets in the first part are arranged in an
approximate order of difficulty for ease of access. It is not intended that students should
work through them one after another, but that they should select those which provide the
practice they need.

The worksheets cover many of the skills specified in the units and elements in the City and
Guilds Numberpower Certificate or, in Scotland, the SQ Numberstart Modules. They should
also prove useful for the equivalent stages of the National Curriculum and NVQ Key Skills.
Many of the worksheets span more than one level and, for entry level work in particular, it
may be necessary to select some questions and omit others on a given sheet.

Detailed answers are provided, except when an individual response is required from the
student or when the answer seems clear. The index *(p. 48)* gives more detailed guidance as to
which worksheets contain practice in specific skills.

Second edition

This second edition of *Maths Worksheets* has been updated, where necessary, mainly to
reflect current prices, statistics, and weights and measures. The information given in *Bar
Charts (p. 26)* remains unchanged because we have been unable to obtain complete up-to-
date figures.

Please read the copyright / photocopying restrictions below.

Publishers: Brown and Brown,
                Keeper's Cottage,
                Westward,
                Wigton
                Cumbria CA7 8NQ
                Tel. 016973 42915

First published 1996
Reprinted 1998

Second edition 2001
Reprinted 2003

ISBN 1 870596 80 3

Other basic maths publications by Brown and Brown

Everyday Maths
ISBN 1 870596 73 0   A5   48 pages
A workbook of basic maths exercises on everyday maths problems and situations.

Everyday Spelling
ISBN 1 870596 83 8   A5   64 pages
A workbook on spelling which includes several pages of exercises on writing numbers
in words as well as practice in writing cheques.

News Worksheet
ISSN 1362-8267   A4   10 pages   Published 3 times per year
Topical worksheets for photocopying, covering reading, spelling, writing and basic maths.

For a copy of the current catalogue, please contact:
Brown and Brown, Keeper's Cottage, Westward, Wigton, Cumbria CA7 8NQ
Tel. 016973 42915

Printed by Reed's Ltd., Penrith, Cumbria on 100% recycled paper and card.

Contents

Place value - Numbers

Numbers must be lined up correctly when adding and subtracting.

e.g. To add 12 and 2 together:

12		12
2		2
14 is correct		32 is wrong

A. *Put the numbers written below into the table on the right, with each figure in its correct place.*

The first one is done for you.

Ten millions	Millions	Hundred thousands	Ten thousands	Thousands	Hundreds	Tens	Units
					1	7	9

one hundred and seventy nine

twenty three

one hundred and twelve

fifty six

five

one thousand, four hundred and eighty two

ten thousand and seventy

six hundred thousand, two hundred and forty one

seventeen million, two hundred and twenty eight thousand, six hundred and ninety one

B. *List these numbers on a separate piece of paper, lining them up correctly under each other.*

| 1,726 | 24 | 22,473 | 600,018 | 92,146,871 | 1 | 695 |

C. *Write these numbers in a list, lining them up correctly under each other.*

two hundred and seventy five

six thousand, nine hundred and twenty eight

six

two million, six hundred thousand

seventeen million and thirty seven

thirty

nine thousand, three hundred

fourteen thousand, six hundred and five

Place value - Money

When adding and subtracting money, the pounds and the pence must be lined up in the correct columns.

e.g. To add £4.60 and 23p together:

4.60	4.60
.23	23.00
4.83 is correct	27.60 is wrong

A. *Fill in the table on the right with the amounts of money in their correct places.*

The first one is done for you.

Thousand pounds	Humdred pounds	Ten pounds	Pounds	Tens of pence	Pence
		2	4	3	3

twenty four pounds thirty three pence

seven pounds ninety two pence

five hundred and twenty six pounds fifteen pence

eight thousand pounds

six hundred and seventy pounds two pence

six hundred and seven pounds

two pence

B. *List these amounts on a separate piece of paper, lining up the pounds and pence correctly.*

£10.99   £1.50   £242.78   .45   £25,760   £1,500   £433.05

C. *Write these amounts in a list, lining up the pounds and pence correctly.*

six pounds, twenty pence

thirty two pounds, fifty pence

two hundred thousand pounds

eighteen pounds, fourteen pence

twenty five pence

one thousand two hundred and seventy two pounds, ten pence

seven hundred and thirty eight pounds, ninety nine pence

six pence

Adding and Subtracting - Figures

A. A van driver has to record her daily mileage in a log book. She records her mileometer reading at the beginning and end of each day and subtracts one from the other to get the day's mileage.

In the chart below, fill in each day's mileage and total them to give the weekly mileage.

Week ending	Mileage at start	Mileage at finish	Mileage
Monday	43,784	43,973	
Tuesday	43,973	44,189	
Wednesday	44,189	44,316	
Thursday	44,316	44,571	
Friday	44,571	44,702	
		Weekly mileage	

B. Complete this pay slip

1. Add up the overtime and the basic wages in the right hand column to give the gross earnings.

2. Add up the deductions *(Pension, National Insurance & Tax)* in the left hand column.

3. Write the deductions total in the right hand column and subtract it from the gross earnings to give the net pay.

NAME... T.R.SILVA ...... W/E.. 7ᵗʰ DECEMBER ....

WORKS/DEPT. No.. A7416 .... Code No... 438L ....

Tax Week No. 36 ....

GROSS WAGES TO DATE		TAX DEDUCTED TO DATE	
£	p	£	p
10442	86	1446	23

WAGES DUE FOR WEEK

		£	p
4 HRS. O/T @ 9·48		37	92
HRS. O/T @			
HRS. O/T @			

DEDUCTIONS

	£	p
Company Pension. 3 %.	7	11
INCOME TAX.........		
National Insurance	19	85
Standard Rate at .22. %	33	92
Reduced Rate at.. 10. %	2	92

OTHER ....

BONUS, HOLIDAY,
SICK PAY, S.S.P. S.M.P.

	£	p
BASIC	237	00
GROSS		

OTHER DEDUCTIONS.. ...
....
....
....

DEDUCTIONS

INCOME TAX REFUND

NET £

Dates and Calendars

A. How many ?

1. How many days in a week ?
2. How many days in a month ?
3. How many weeks in a year ?
4. How many months in a year ?
5. How many days in a year ?
6. How often does a leap year occur ?

B. Number the months

The date is often written in numbers *e.g.* 9/4/05 **or** | 0 9 | 0 4 | 0 5 |

1. a. Which month does the **4** stand for ?
   b. How would the above date be written in the U.S.A. ?

2. Which months are these ?

   3   11   5   9   12   1   7   2   10   6   8

C. Using a calendar

Answer the questions about this calendar for August 2002.

	August					
M	Tu	W	Th	F	Sa	Su
			1	2	3	4
5	6	7	8	9	10	11
12	13	14	15	16	17	18
19	20	21	22	23	24	25
26	27	28	29	30	31	

1. A birthday is on 14th August. What day of the week will it be ?
2. How many Saturdays are there in the month ?
3. What is the date of the first Sunday in the month ?
4. What is the date of the last Thursday in the month ?
5. How many days are there between Friday 16th and Thursday 22nd ?
6. What day of the week will September 2nd 2002 be ?
7. What day of the week will July 28th 2002 be ?
8. A 4-day weekend break in Paris is advertised. It leaves on Friday 9th and returns on Monday 12th August. How many nights will there be in Paris ?
9. a. Why is August 26th highlighted ?
   b. Why would this calendar not be useful in Scotland or the Republic of Ireland ?

D. In your opinion

1. Which are the summer months ?
2. Which months are in Autumn ?
3. Which are the winter months ?
4. Which months are in Spring ?

Note: The answers may depend partly on where you live !

Time and Clocks (1)

A. How many ?

1. How many minutes in an hour ?

2. How many hours in a day ?

3. How many seconds in a minute ?

4. How many hours are there *a.m.* and how many hours *p.m.* ?

5. How many times a year are clocks changed in the U.K. ?

B. Time

> *The time can be expressed in 3 ways:*
> * In words   *e.g. It is a quarter to eight*
> * On the 12-hour clock   *e.g. 7.45 a.m.* **or** *7.45 p.m.*
> * On the 24-hour clock   *e.g. 0745 hours* **or** *1945 hours*

1. *Re-write each of these times as a 12-hour clock number and a 24-hour clock number.*
   e.g. eight o'clock in the morning = 8 a.m. *(12-hr clock);* 0800 hrs *(24-hr clock)*

	12-hr	*24-hr*
half past ten in the morning		
ten to two in the morning		
a quarter past seven in the evening		
a quarter to ten in the evening		
noon		
two minutes past eleven in the morning		
five to two in the afternoon		
twenty to four in the afternoon		
midnight		
seventeen minutes to six in the morning		

2. *Re-write these 24-hour clock times as 12-hour clock times.*
   e.g. 1723 = 5.23 p.m.

0730	1640
1122	1200
2306	0001
1853	2115

   How would you give each of the above times in words ?

Time and Clocks (2)

A. Start and Finish

Write down the answers to these questions as a 12-hour clock time and a 24-hour clock time.

1. A recipe gives a cooking time of 1 ¹/₂ hours. You want the meal to be ready by eight o'clock in the evening and the oven needs to heat up for 15 minutes. What time do you set the oven timer to start and to finish ?

2. Another recipe gives a cooking time of 50 minutes and the oven needs 20 minutes to heat up. If you want the dish to be ready at a quarter to one in the daytime, what times should you set the oven timer to start and to finish ?

3. You are boiling a bacon joint for the main meal of the day. The joint weighs 1.75kg and the cooking time is calculated at 25 minutes per 500g plus 20minutes extra at the end, after the water reaches boiling point. Decide what time you want the meal, then work out when you should start cooking the bacon.

4. An electronic timer on a central heating system is set to come on for two hours at a quarter to seven in the morning. It comes on again in the evening for six and a half hours until midnight. To what times is the timer set ?

5. You want to set a video recorder to record a 2-hour TV programme starting at ten minutes to midnight. What are the start and stop times ?

6. You set an electronic timer switch to turn on some lights while you are out in the evening. It is starting to go dark at about six and you want the lights to stay on until about one in the morning. To what times do you set the timer ?

7. Use a copy of the TV & Radio programme guide from today's newspaper to do these:

    a. Pick a film which you could video and decide on the start and stop times.

    b. Your sister is getting married tomorrow, 100 miles away. You want to record the most useful weather forecast. Decide which one to record and the start and stop times you would use.

    c. There has been flooding in your area and you want to record the local news and weather programmes throughout the day. Make a list of the programmes and their start and stop times.

B. Planning a journey

You have to go to an interview in a large city over 200 miles away. Plan your journey using mainly public transport. Make a list of the times for each stage of the journey from leaving home to returning.

C. Planning a charity fund-raising day

Write out the timetable for a fund-raising day for *Children in Need*.

Adding and Subtracting - Hours

A. A postman works six days a week and his basic day is from 5.30 a.m. to 12.30 p.m. If he works after 12.30 p.m., he is paid overtime. He fills in his time sheet at the end of each day. At the end of the week, he has to add up all the overtime in hours and minutes, then round it to the nearest number of hours.

On the time sheet below:

1. Fill in the overtime column.
2. Add up the overtime to give a total in hours and minutes.
3. Round the total to the nearest number of hours.

	Start	Finish	Overtime hrs	Overtime mins
Monday	5.30	12.45		
Tuesday	5.30	12.30		
Wednesday	5.30	1.45		
Thursday	5.30	12.55		
Friday	5.30	3.15		
Saturday	5.30	12.30		
Total overtime				
Overtime paid				hours

B. An electrician, who is self-employed, works as a sub-contractor for a builder. Each day he has to record the hours *(to the nearest ¼ hour)* that he works on 5 different contracts. At the end of the week, he totals up the hours on each job.

On the time sheet below:

1. Add up the hours spent on each job.
2. Add up the hours to give the weekly total.

Week ending: 14th Jan.

Job title	Mon	Tues	Weds	Thurs	Fri	Sat	TOTAL
Prestwick Airport	5 ¼		1 ½				
Co-op Store	¾	2	1 ½				
Girvan Road	2 ¼	¾			6 ½	4 ¾	
Craiglee Hotel		3	4 ¾	8			
Shepherd & Watson		2 ¼			2	¾	
						WEEKLY TOTAL	

Multiplication and Division

Multiplication

You should be able to multiply numbers up to 10 x 10 in your head. This table shows the multiplication of numbers up to 10 x 10.

1	2	3	4	5	6	7	8	9	10
2	4	6	8	10	12	14	16	18	20
3	6	9	12	15	18	21	24	27	30
4	8	12	16	20	24	28	32	36	40
5	10	15	20	25	30	35	40	45	50
6	12	18	24	30	36	42	48	54	60
7	14	21	28	35	42	49	56	63	70
8	16	24	32	40	48	56	64	72	80
9	18	27	36	45	54	63	72	81	90
10	20	30	40	50	60	70	80	90	100

There is no quick way to learn multiplication tables. It is best to try to learn one line at a time.

Learning tables

Start with the **2 x** *line:*

a. Say it through.

b. Write it down.

c. Cover it up.

d. Say it through again.

e. Write it down again.

Repeat **a.** *to* **e.** *over a few days until you are sure of it.*

Try a few exercises on each line as you learn it. There are 2 examples in the boxes below.

Fill in the missing numbers in this line
2   4   __   8   10   __   14   __   18   __

Fill in the answers		
7 x 2 =	9 x 2 =	4 x 2 =
10 x 2 =	3 x 2 =	5 x 2 =
8 x 2 =	2 x 2 =	6 x 2 =

Repeat the above method for each line of the multiplication table.

Division

Some simple division can be done in your head but it is best to use a calculator for more difficult division. The multiplication table above can also be used for division.

e.g.   A pack of 5 nuts and bolts costs 45p. You want to know the cost of 1 nut and bolt.

Look along the **5** *line until you get to* **45**. *Look up to the top of that column and the answer is* **9**. *The cost of 1 nut and bolt is* **9p**.

Calculating the area

A.
1. A building plot for a house measures 16.5 metres by 24.7 metres. What is the area of the plot in square metres ?

2. In imperial units the same plot measures 18 yards by 27 yards. What is the area of the plot in square yards ?

3. Which is larger - a square metre or a square yard ?

B. A corridor is 2.0 metres wide and 9.5 metres long.
1. What is the floor area of the corridor ?
2. How many carpet tiles *(500 mm x 500 mm)* will be needed for the corridor ?

C. The owner of a house needs to repaint the 4 outside walls. The 2 side walls measure 8.8 m x 5.1 m and the end walls measure 5.5 m x 5.7 m *(average height)*.

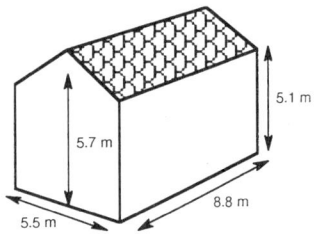

1. Calculate the total area of the 4 walls.

2. How many 5 litre tins of paint will be needed to give 1 coat to all the walls at the recommended spreading rate of 8 m² per litre ?

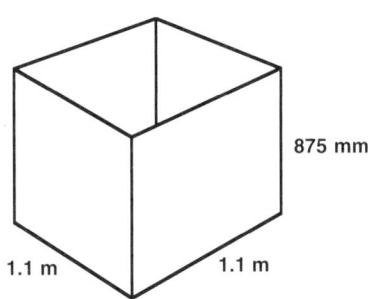

D. A compost bin is to be made from old floorboards. The finished bin will measure 1.1 m x 1.1 m x 875 mm high.

1. What will be the area of one side of the bin in square metres ?

2. How many boards will be needed to make the four sides of the bin, if the floorboards measure 22 mm x 125 mm x 2.5 m ?

E. A DIY enthusiast offers to build a raised flower bed for a disabled man. The measurements of the bed are 3 m x 1.2 m x 0.5 m high.

1. What is the area of each of the long side walls ?
2. What is the area of each of the end walls ?
3. If 60 bricks are needed per square metre, how many will be needed for the whole flower bed ?

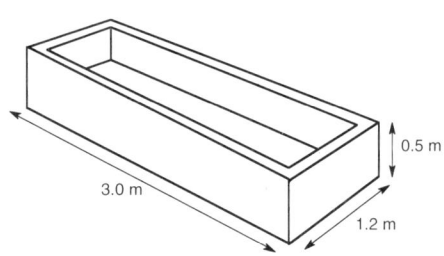

F. A burglar alarm sensor claims to cover an area of about 28 m². Will it be able to cover all the floor area of these ?
1. A workshop which measures 5 m x 5 m ?
2. A workshop which measures 3.2 m x 7.1 m ?

Calculating the volume

For each of the questions on this page:

1. *Write down a rough estimate in pencil.*
2. *Use a calculator to work out the exact answer and write it in ink.*

A. Volume of rooms

A plumber has to work out what size of radiator is needed for each room in a house. In order to decide on the size, he has to measure the volume of each room.

Volume = length x width x height

The measurements of the rooms in metres are given below. Work out the volume of each room.

1.	Hall	2.4 x 2 x 2.4
2.	Lounge	5 x 4 x 2.4
3.	Dining room	3 x 3 x 2.4
4.	Kitchen	2.7 x 2.4 x 2.4
5.	Double Bedroom	3.6 x 3 x 2.4
6.	Single bedroom	2.7 x 2.7 x 2.4
7.	Bathroom	2.4 x 2 x 2.4

B. Volume of an oil tank

A family moves into an old house. There is an oil tank for the central heating system but nothing to say how much oil it holds.

The tank measures: 1.5 m x 1.1 m x 1.3 m

1. What is the maximum volume of the tank in cubic metres ?

2. *There are 1000 litres in a cubic metre.* What is the capacity of the tank in litres ?

C. Volume of a cold water tank

A cold water tank measures 62 cm long and 42 cm wide. The maximum depth of water it holds is 45 cm.

1. *1 litre = 1000 cubic centimetres.* What is the maximum capacity of the tank in litres ?

2. What is the volume of the tank in cubic metres ?

D. Volume of a raised flower bed *(see Calculating the area: p.12, Section E)*

The outside measurements of the bed are: 3 m x 1.2 m x 0.5 m.
The brick walls are 100 mm thick.

Work out the maximum volume of soil that the raised flower bed will hold.

Fractions

A fraction is usually **written** with one number over the other: $\frac{1}{2}$
but these days it is usually **printed** as 1/2

This often makes it difficult to see what, for example, 1 1/2 or 3 3/4 means

In *Maths Worksheets*, all fractions are printed this way: $^1/_2$ $^1/_4$ $1^1/_2$ $3\,^3/_4$

A. Alongside each of these words, write down the fraction in figures:

one eighth	one quarter	one third
three-eighths	a half	five-eighths
two-thirds	three-quarters	seven-eighths
one and a quarter	nine and three quarters	three and a half

B. Look at these diagrams and write down in figures for each one:

● the fraction that is shaded   ● the fraction that is not shaded

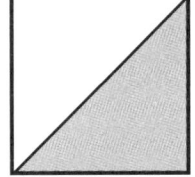  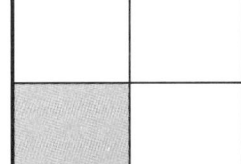 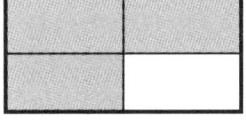 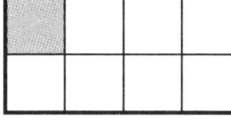

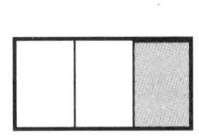 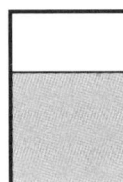 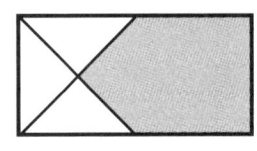 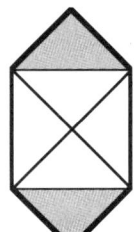 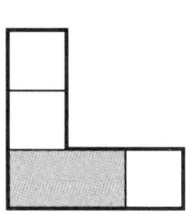

C. Shade in the fractions asked for alongside each shape.

$^3/_4$  $^1/_2$ 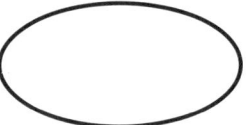 $^1/_4$

$^3/_8$ 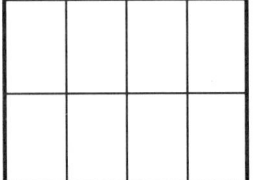 $^7/_8$ 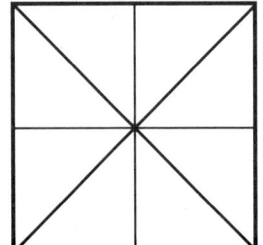 $^5/_8$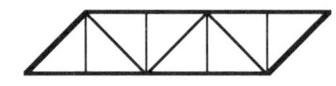

Proportion

News reports of surveys or opinion polls are often given as a proportion.
A proportion is just a fraction by another name. For example, a news report may say:
Three out of every four households now have a video.   **3** out of **4** is the same as $3/4$

A. Choose a proportion from the box to fit the gaps in each of these statements:

1 in 100
1 in 12
1 out of 10
1 in 7
1 in 5
3 out of 4
1 in 2
2 in 3
7 out of 10

☐   ___ in every ___ marriages in Britain now ends in divorce.

☐   ___ out of ___ apples sold in the U.K. are imported.

☐   ___ out of ___ people in the U.K. are vegetarians.

☐   ___ in every ___ small businesses closes down each year.

B. What proportion is female and what proportion is wearing a hat ? *(Not including the dog!)*

C. What proportion points: **1.** to the right ?   **2.** to the left ?

D. The compass on the right shows the direction of **North**.

Write down the proportion of the arrows which points:

1. East   **3.** South

2. West   **4.** North

E. 1. What proportion of these is **b** ?

d d b d d d b d b b

2. What proportion of these is smiling ?

Decimals

A. Common decimals

It is very useful to be able to remember common fractions as decimals.

Fill in the decimal value of each of these:

$1/2 =$ $1/4 =$ $1/3 =$

$1/5 =$ $1/10 =$ $1/100 =$

B. Money

Money is often written in decimal form. The important thing is to put the decimal point in the right place. When writing pence as a decimal of a pound, *always* put 2 figures after the decimal point.

Examples: **£40** is written as **£40.00**

                            50p is written as **£0.50** *never* as **£0.5**

                            3p is written as **£0.03**

Write each of these amounts in decimal form. Arrange each group in a column and add it up:

1. £100   £50   £8   90p   2p

2. £120   £75.60   50p   25p   3p

3. £8.99   £172.10   £18   £1   64p

4. £143.75   £1   58p   74p   6p

C. Multiplying decimals by 10 or 100

- To multiply a decimal number by **10**, move the decimal point **1 place to the right**.

         e.g. $36.25 \times 10 = 362.5$

                   $0.017 \times 10 = 0.17$

- To multiply a decimal number by **100**, move the decimal point **2 places to the right**.

         e.g. $84.361 \times 100 = 8436.1$

                   $0.017 \times 100 = 1.7$

Fill in the answers to these:

$6.27 \times 10 =$ $0.317 \times 10 =$

$18.90 \times 100 =$ £3.78 \times 10 =$

£0.52 \times 100 =$ 38p \times 100 =$

$7.48 \text{ m}^2 \times 10 =$ $967.1 \text{ cm} \times 100 =$

$46.01\text{g} \times 100 =$ $78{,}010.01 \times 100 =$

$5{,}496.202 \times 10 =$ £183{,}642.79 \times 100 =$

Percentages

A percentage is a special type of fraction. It is the fraction of 100.

The sign **%** means *out of 100*

Examples

❏ **37%** means **37 out of 100**. It could be written as the fraction $^{37}/_{100}$

❏ In the diagram on the right, there are **100** squares:

12 of the **100** squares are shaded

i.e. **12%** of the squares are shaded

❏ Think of a percentage as if it were pence in a pound.

e.g. To imagine what **64%** is, think of **64p** as a proportion of **£1**.

A. *Write down each of these fractions as a percentage:*

$^{16}/_{100}$ $^{72}/_{100}$ $^{45}/_{100}$ $^{8}/_{100}$ $^{99}/_{100}$

B. *Write down each of these as a percentage:*

1. 3 in every 100

2. 12p in every £1

3. 87p in every £1

4. 100 in every 200

C. *Percentages are very common in everyday life. Many facts and figures are expressed in percentages and they are also used in common financial terms. For example, interest rates, increases in pay, rents or charges are usually stated as a percentage. VAT is always calculated as a percentage* (see *Calculating VAT: p. 44*).

True / False / Don't know

1. An employee who is offered a pay increase of 3% above inflation is likely to accept it.

2. A football manager who asks his team to give 150% is asking the impossible.

3. A building society that pays investors 5.5% might charge mortgage borrowers 8%.

4. A loan company that charges credit at an APR of 85% is ripping people off.

5. A personal loan from a bank might have an APR of 18% but in a credit arrangement with a shop the same sum might have an APR of 30%.

6. If inflation in the U.K. was running at 10%, it would be at a high level.

7. A restaurant that has a 20% service charge is charging customers too much.

Rounding numbers

Rounding numbers up or down to a handy figure is an important skill in maths.

A.

> **Rounding to the nearest 10**
>
> Numbers which **end** in **1, 2, 3, 4** are rounded **down**
> Numbers which **end** in **5, 6, 7, 8, 9** are rounded **up**
> Numbers which **end** in **0** are already rounded
>
> *e.g.* 84 is rounded down to 80
> 87 is rounded up to 90
> 85 is rounded up to 90
> 80 is already rounded

Round each of these figures to the nearest **10**:

89   41   156   32   477   13   1,968   524   65   100   95

B.

> **Rounding to the nearest 100**
>
> Numbers in which the **tens** figure is **1, 2, 3, 4,** are rounded **down**
> Numbers in which the **tens** figure is **5, 6, 7, 8, 9** are rounded **up**
> Numbers which **end** in **00** are already rounded
>
> *e.g.* 334 is rounded down to 300
> 374 is rounded up to 400
> 354 is rounded up to 400
> 300 is already rounded

Round each of these figures to the nearest **100**:

123   594   651   278   449   933   824   199   510   1,750

C.

> **Rounding decimals**
>
> To round to the nearest **whole number,** use the first figure after the decimal point.
>
> *e.g.* 6.4   is rounded down to 6
> 6.5   is rounded up to 7
> 7.946 is rounded up to 8 (ignore the 46)
> 6.0   is already rounded

Round each of these figures to the nearest **whole number**:

8.6   17.2   10.9   3.1   94.7   165.3   99.8   0.48   7.5   108.09

D. Round each of these figures to the nearest **0.1**:

e.g. **3.77** rounds up to **3.8**   **3.74** rounds down to **3.7**

1.36   14.55   7.14   271.61   3.23   8.45   18.95   19.305   111.777

Rounding numbers on a calculator

- **To round up or down on a calculator, ignore all the unnecessary decimal places.**

  e.g. On a calculator, 22 divided by 7 = **3.1428571**

  To round **3.1428571** to the nearest whole number: Ignore 428571
                                                                    Round 3.1 down to 3

  To round **3.1428571** to 2 decimal places: Ignore 8571
                                                                 Round 3.142 down to 3.14

- **When using a calculator to work out money, the same rules apply.**

  e.g. On a calculator, £90 divided by 7 is: 90 ÷ 7 = **12.857142**

  To round **12.857142** to 2 decimal places: Ignore 142
                                                             Round 12.857 up to 12.86 *i.e.* **£12.86**

Note: *It often helps to cover up the unnecessary figures on the calculator display with a finger, so that it is easier to see the important figures.*

A. *Round these figures to exact pounds and pence:*

    £1.654   £684.9687   £ 7.995   £1324.6432   £0.33333   4.8p   95.5p

B. *Using a calculator, work out each of these. Write the answer in figures* (pounds & pence).

1. If six people win £2500 on the Pools, how much does each person get ?

2. A bill for £19 is to be shared by three people. How much will each person pay ?

3. What is a quarter of £222 ?

4. What is a seventh of £498 ?

5. What is a fifth of £65 ?

6. A pack of 24 Christmas cards costs £2.85. What is the cost of each card, to the nearest 1p ?

7. A mail order firm asks customers to add 15% to all orders under £50 for post and packing. Orders over £50 are post free. How much will need to be added for orders costing the amounts below ?

       a. £17.75   **b.** £48.99   **c.** £103.31

8. A wholesaler offers 5% off its published prices to customers who pay cash with order. How much would customers save and how much would they have to pay for orders of these values ?

       a. £178.95   **b.** £2589.77   **c.** £364.11

9. A bookshop pays two-thirds of the selling price for the books it buys from a publisher. If the selling price of a book is £14.99, what is the cost to the bookshop of:

       a. 1 book ?   **b.** 25 books ?

Estimating (1)

One of the most important and useful of all mathematical skills is that of estimating. Calculators don't make mistakes but the humans who press the buttons do ! So it is important to be able to estimate what an answer should be. It is also very useful to be able to make rough estimates in your head without having to measure things exactly and work out the sums on paper.

Example: A woman sees some bargain towels for sale at £3.85 each. She only has £10. How many towels can she buy ?

Rough estimate:	Each towel is about £4.
	2 towels would be about £8.
	3 towels would be about £12.
Answer:	She can only buy two.
Accurate calculation:	2 towels @ £3.85 = £7.70
	3 towels @ £3.85 = £11.55
Answer:	She can still only buy two !

For each of the questions below:

A. *Make a rough estimate of the answer.*   **B.** *Work out the exact answer - if there is one !*

1. What will be the cost of a dozen 19p stamps ?

2. Potatoes are for sale at 3 kg for 95p or 10 kg for £2.50. Is the bigger bag worth it ?

3. A journey consists of a train journey of 2 hours 48 minutes and a bus ride of 50 minutes. If you leave on the 10.30 a.m. train, what is the earliest you can expect to arrive ?

4. A farmer sells a cow at an auction. The price is £1.19 pence per kilogram and the cow weighs 521 kg. How much will he get ?

5. A salesman selling office furniture gets a commission of £4.80 on every filing cabinet sold. Last month he sold 37. How much commission will he get ?

6. A haulage firm has to collect 21 castings from a factory. Each weighs 1.6 tonnes. What will the total weight be ?

7. A kitchen work-top is 3 metres long. Will it fit inside a small hatchback car or will it have to go on a roof rack ?

8. A step ladder is 1.9 metres high. Will it enable someone to reach a smoke alarm on a ceiling 4.5 metres high ?

9. A television costs £299.99 to buy. On credit, it costs £12 per month for 31 months plus a £30 deposit. How much more does it cost on credit ?

10. What is the height of the tree ?

Estimating (2)

For each of the questions below:

A. *Make a rough estimate of the answer.*

B. *Work out the exact answer - if there is one !*

1. How long is it going to take you to answer all the questions on this sheet ?

2. A restaurant meal costs £58.52 for seven people. What is the cost per person ?

3. A pack of 5 dust bags for a vacuum cleaner costs £3.25. What is the cost per bag ?

4. **Rangers**   *Played* **34**   *Points* **65**
   Celtic   *Played* **32**   *Points* **58**

   Rangers have 2 games left to play but Celtic have 4 games left. Have Celtic got much of a chance of winning the League ?
   (3 points for a win; 1 point for a draw)

5. A package holiday in a hotel in Spain costs £319 for a week. 2 weeks cost £444. What will it cost for 3 weeks ?

6. If you were going to your grandmother's funeral 100 miles away and you had to leave home by 7 a.m., for what time would you set your alarm ?

7. What time would you leave home to do any of these by 9 a.m. ?

   a. get to work   **b.** get to college/school   **c.** sign on

   d. get children to school/play group/nursery

8. **a.** A long distance 'phone call at peak rate costs 8.1p per minute. How much will a 6 1/2 minute call cost ?

   b. How much will the same call cost on a mobile 'phone on which the charge is 23p per minute ?

9. How long will it take you to copy 100 addresses on to envelopes, put a letter into each envelope, lick the envelopes and put on stamps ?

10. How much coffee, milk and sugar and how many packets of biscuits will be needed for a coffee break in a training course for 50 people ?

Did you know ?

If a builder is asked to give an *estimate* for a job (*e.g.* a house extension), he will probably give a figure to the nearest £500 or £1000. For example, he might give an *estimated* figure of £12,500. The final price of the job would probably not be exactly that amount.

If a builder is asked for a *quotation* for a job, he should give an exact figure, based on detailed drawings, for which he will agree to do the job.

Rulers and Measuring Tapes (1)

A. *This ruler is in millimetres (mm).*
   ***35 mm** is marked*

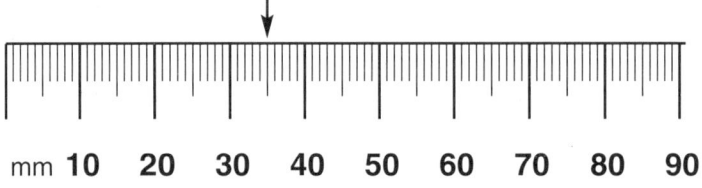

Mark these points on the ruler: 20 mm   54 mm   47 mm   3 mm

B. *This ruler is in centimetres (cm).*
   ***1.5 cm** is marked*

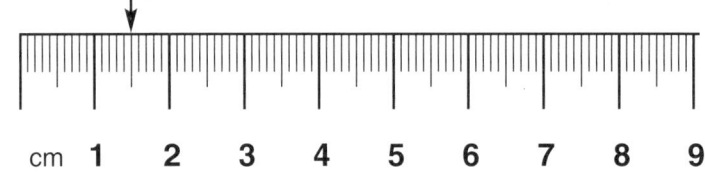

Mark these points on the ruler: 3 cm   5.9 cm   4.2 cm   0.2 cm

C. *This ruler is in inches (").*   **2 ¹/₂ "** is marked*

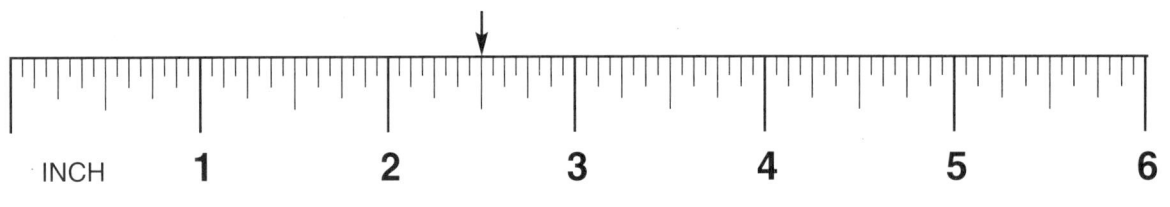

Mark these points on the ruler: 3 "   3 ³/₄"   1 ⁵/₈"   ⁷/₈"   ¹/₄"

D. *This steel measuring tape is in both centimetres and inches. The section shown is around the*
   ***100 cm (40")** mark.*

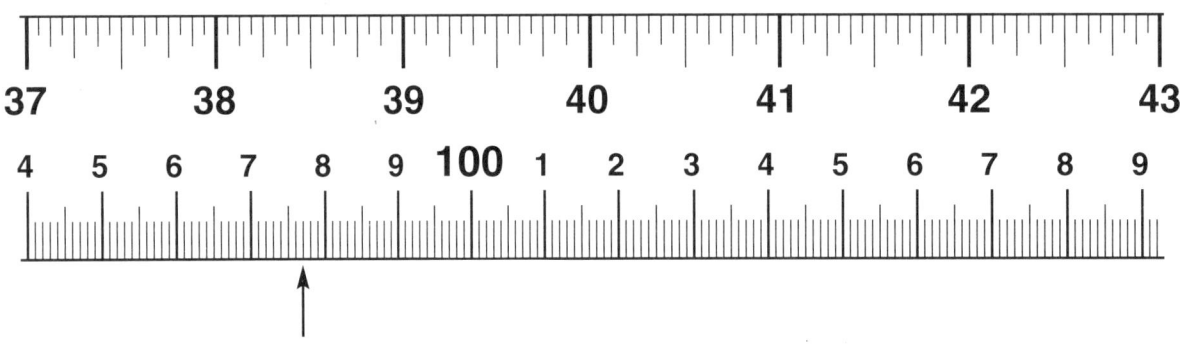

1.   **97.7 cm** is marked. What is the equivalent length in inches ?

2.   Mark these on the tape:
     105 cm   992 mm   1.03 m   38 ³/₄"   40 ⁷/₈"   41 ¹/₈"

Rulers and Measuring Tapes (2)

A. *What is the length of this line ?*

_____

mm: **ins:**

B. *What is the height and width of a piece of A4 paper ?*

mm: **cm:**   **ins:**

C. *On an A4 piece of paper, draw lines of the following lengths:*

192 mm   6.3 cm   7 1/4 "   4 mm   23 cm   10 5/8 "   351 mm

D. *Using a steel tape, measure the size of the desk or table you are writing on in each of these:*

mm: **cm:**   **metres:**

inches: **feet and inches:**

E. *Using a steel tape:*

   1.   Measure out the lengths **194.5 cm** and **69 7/8 "** on the floor.

   2.   Measure the distance round the inside of a door frame:

   mm: *ins:*

   3.   Measure the distance from the floor to the underside of a table or desk:

   mm: *ins:*

   4.   Measure the distance between 2 shelves:

   mm: *ins:*

   Note: *Read the distance to the inside edge of the tape box*
   and then add 50 mm for the length of the tape box.
   (For the measurements in inches, add 2")

   + 50mm
   Measurement
   in mm

F. *Using a soft dressmaking tape, measure:*

   Your bust/chest: *cm:*   *ins:*

   Your waist: *cm:*   *ins:*

   Your hips: *cm:*   *ins:*

   Your inside leg length: *cm:*   *ins:*

   Your neck *(collar size)*: *cm:*   *ins:*

Scales and Dials

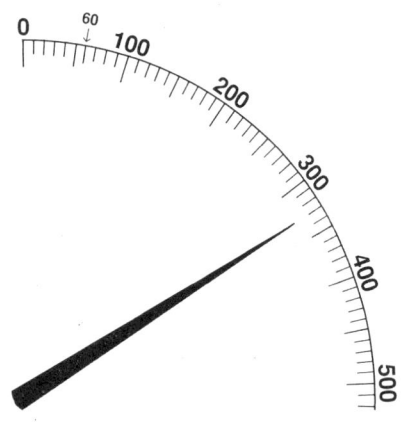

A. *This postal scale for letters is in grams.*

   1.   What weight does the pointer show ?

   2.   Why is **60g** marked separately ?

   3.   Mark these weights on the outside of the scale:

200g	375g	125g
210g	150g	40g

B. *This car speedometer shows the speed in miles per hour (mph).*

   1.   How fast is the car going ?

   2.   Mark the following speeds:

70 mph	35 mph
88 mph	43 mph
12 mph	25 mph

C. *These kitchen scales show the weight in kilograms and grams on the inside, and pounds and ounces on the outside.*

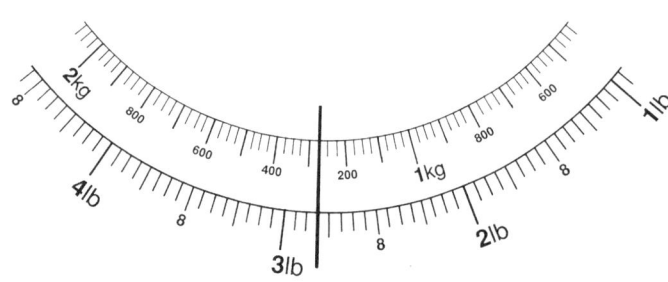

   1.   What weight does the pointer show:

      a. in kg & g ?

      b . in lb & oz ?

   2.   Mark these weights on the scale:

1.5kg	1kg 150g	625g
1 lb 3 oz	3 lb 14 oz	2 $1/2$ lb

D. *These bathroom scales show weight in kilograms and in stones and pounds.*

   1.   What weight is the line showing ?

     ● In kilograms

     ● In stones & pounds

     ● In pounds only

   2.   Mark the following weights:

53 kg	8 st 3 lb
77 kg	11 st 13 lb

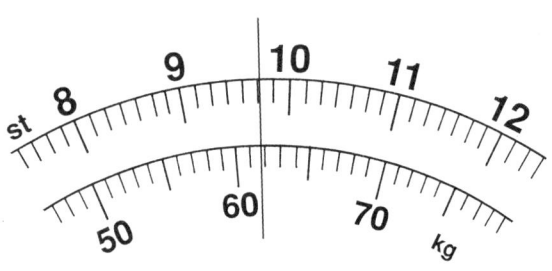

Thermometer and Radio scales

A. Medical thermometer

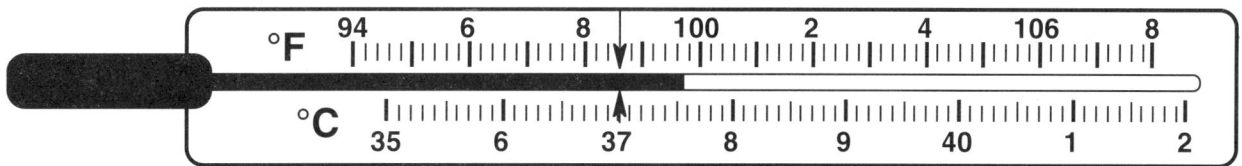

The arrows on the medical thermometer indicate normal body temperature.

Answer each of these questions in °C and in °F

1. What temperature is normal body temperature ?

2. What temperature does the thermometer read ?

3. Mark these temperatures on the thermometer and read off the value of each on the opposite scale: **a.** 101.7°F **b.** 39.3°C

4. What temperature might you expect in someone who is very seriously ill ?

B. Radio frequency bands

This radio has 3 frequency bands: **FM** (was known as **VHF**)

MW (medium wave)

LW (long wave)

*The scale on the **FM** band is regularly spaced, like a ruler.*
*The spacing for the **MW** and **LW** bands is not regular.*

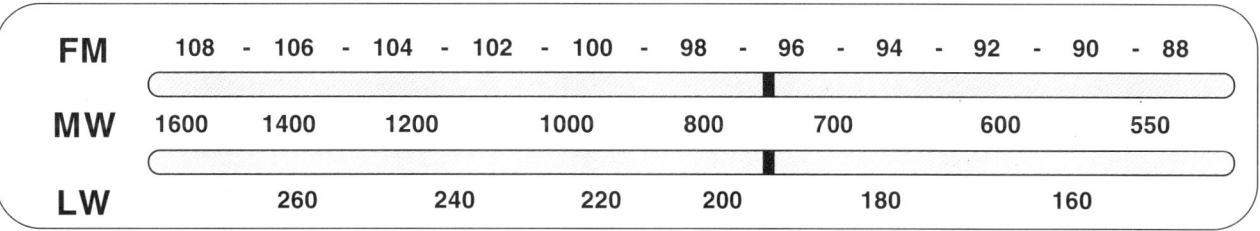

1. What frequency *(roughly)* is the radio shown above tuned to ?
   a. On **FM**   b. On **MW**   c. On **LW**

2. Mark these frequencies on the scale:

BBC Radio 1	**98.5 FM**
BBC Radio 5 Live	**909 MW**
BBC Radio Belfast	**92.4 - 95.4 FM** *or* **1341 MW**
Virgin	**1215 MW**
Atlantic	**252 LW**
Your local radio station	

Bar charts

One of the simplest ways of presenting numerical information is in the form of a bar chart (or bar graph). The bar chart below shows the number of people, out of every 100,000 of the population, who were in prison in 1992.

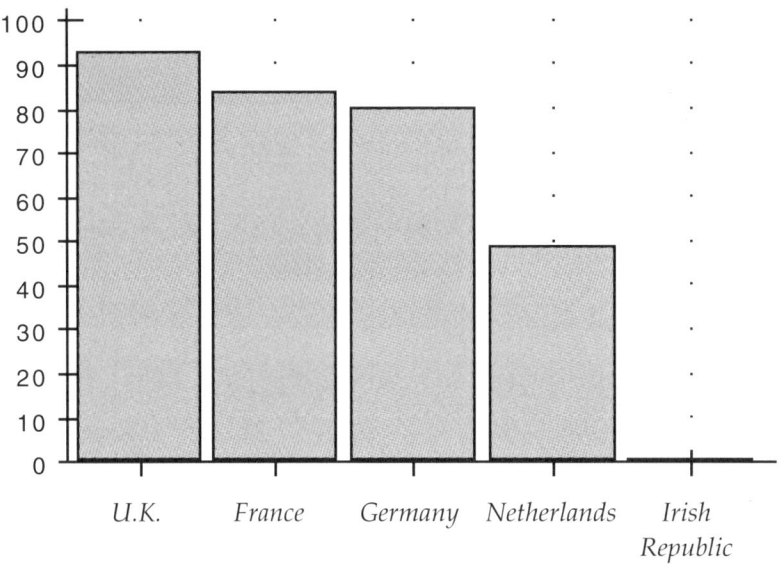

A. The Irish Republic had 62 prisoners per 100,000 population.

Draw in its bar on the chart.

B. True / False / Maybe

    1. In 1992, there were more people in prison in the U.K. than in The Netherlands.

    2. In 1992, there were more people in prison in the Irish Republic than in Germany.

    3. There was more crime in the U.K. in 1992 than in the other 4 countries.

    4. The U.K. is better at catching criminals than the other 4 countries.

    5. The U.K. imposes longer sentences than the other 4 countries.

    6. The U.K. imprisons people for more offences than the other 4 countries.

C. Bar charts can be drawn horizontally as well as vertically.

Using squared paper, re-draw the bar chart above with the bars horizontal.

D. *Draw 2 bar charts (one for men & one for women) using these percentage figures on how people get jobs:*

	Answering adverts etc.	Jobcentre *or* Careers Office	Direct approach to employers	Personal contacts	Other methods
Men	36	34	10	13	7
Women	53	26	7	9	5

This page is about *bar charts*. What is a *bar code* ?

Pie Charts

Pie charts are used to show, at a glance, how things are shared out.

A. This pie chart shows how money is divided by a County Council between different services.

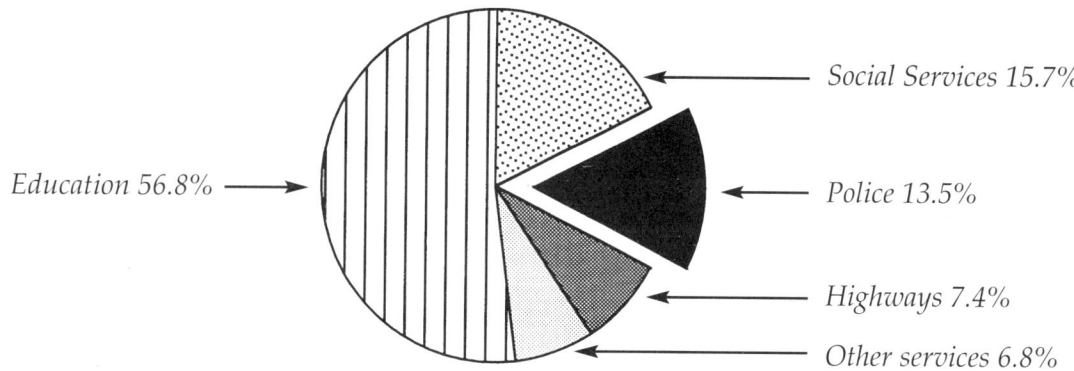

Education 56.8% →

Social Services 15.7%

Police 13.5%

Highways 7.4%

Other services 6.8%

1. Which Council service gets more money than any other ?
2. Does Social Services get a quarter of the total budget ?
3. Is it true to say that Social Services gets over twice as much as Highways ?

B. The two pie charts below show weights of men and women in England compared with the recommended weight for their age and height.

The charts are based on these figures:

Women : *Underweight:* **9%**; *Desirable weight:* **50%**; *Overweight:* **26%**; *Very overweight:* **15%**

Men : *Underweight:* **6%**; *Desirable weight:* **41%**; *Overweight:* **40%**; *Very overweight:* **13%**

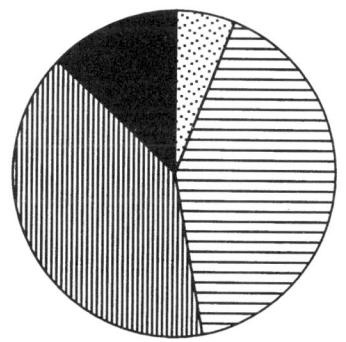
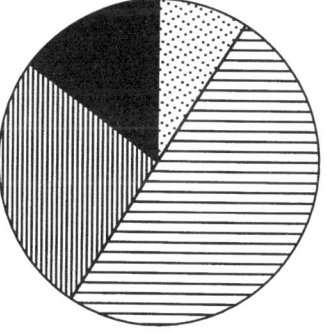

1. Decide which chart is for men and which is for women. Write in the figures for each section.

2. Fill in the gaps in these sentences with either **men** or **women**.

   a. More _____ are overweight than _____ .

   b. More _____ are underweight than _____ .

   c. More _____ are grossly overweight.

   d. Fewer _____ than _____ have the correct body weight.

   e. More than half of all _____ are overweight.

Line graphs

Line graphs show the relationship between two sets of figures.

A. The most common graph gives dates along the horizontal scale and figures for things like prices, profits, goods sold etc. up the vertical scale.

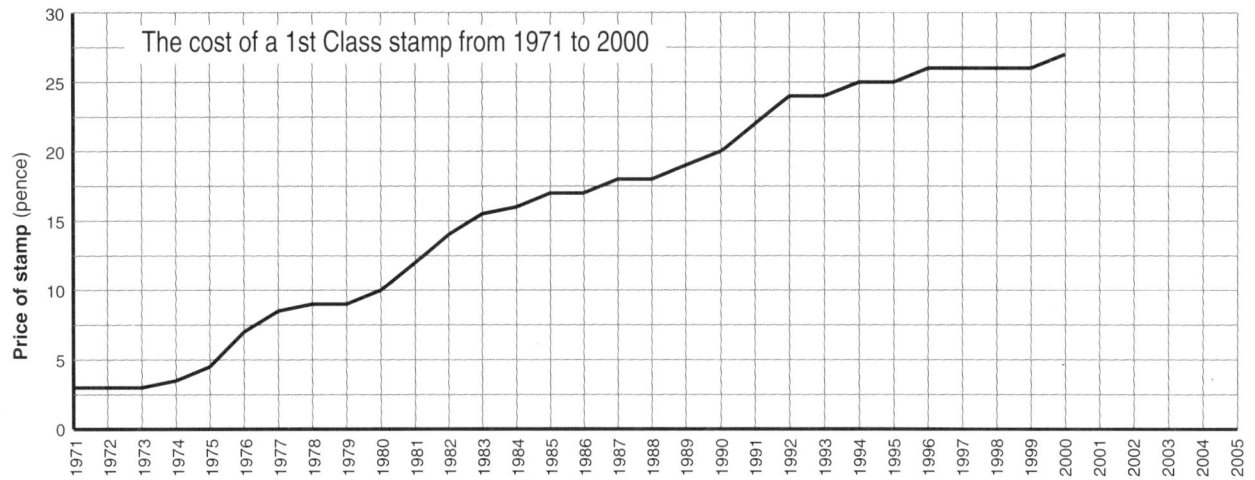

What was the cost of a 1st Class stamp in each of these years ?

1. 1971 **2.** 1984 **3.** 1995 **4.** 2000

B. Graphs can be made to show information in a different light by changing the scale.

Example

U.K. unemployment figures			
	Males	*Females*	*Total*
May 2000	850,000	258,000	
June 2000	825,000	253,000	
July 2000	821,000	268,000	
August 2000	814,000	275,000	
September 2000	785,000	257,000	

If the Government wants to show that unemployment is falling rapidly, it might use the graph on the left. The graph on the right uses the same figures but presents them less dramatically.

U.K. male unemployment *(in thousands)*

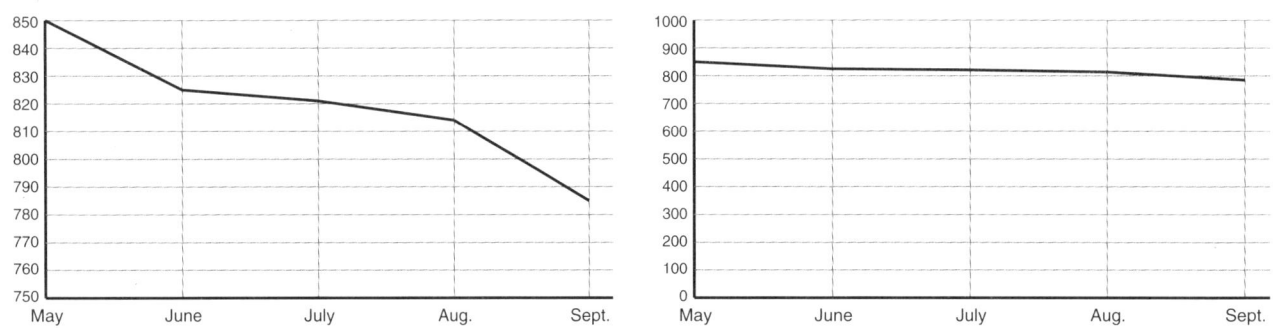

Draw similar graphs to show the female unemployment figures and the total unemployment figures (male + female).

C. *Draw a graph with the 12 months of the year along the horizontal. Draw a line for each of these supermarket items which shows roughly how you think sales might go over a year.*

1. Ice cream **2.** Frozen turkeys **3.** Tins of soup

Maps (1)

All maps, atlases and town plans use a grid system to show where places are. In the index, the horizontal reference is often given first, followed by the vertical reference.

The simplest index gives references to **one square** in the grid. Street maps such as *A-Z guides* use this system. Having found the square, you then have to look in it for the street or place name.

The map below shows the island of Anglesey in North-West Wales.

A1 refers to the bottom left-hand square   **G5** refers to the top right-hand square.

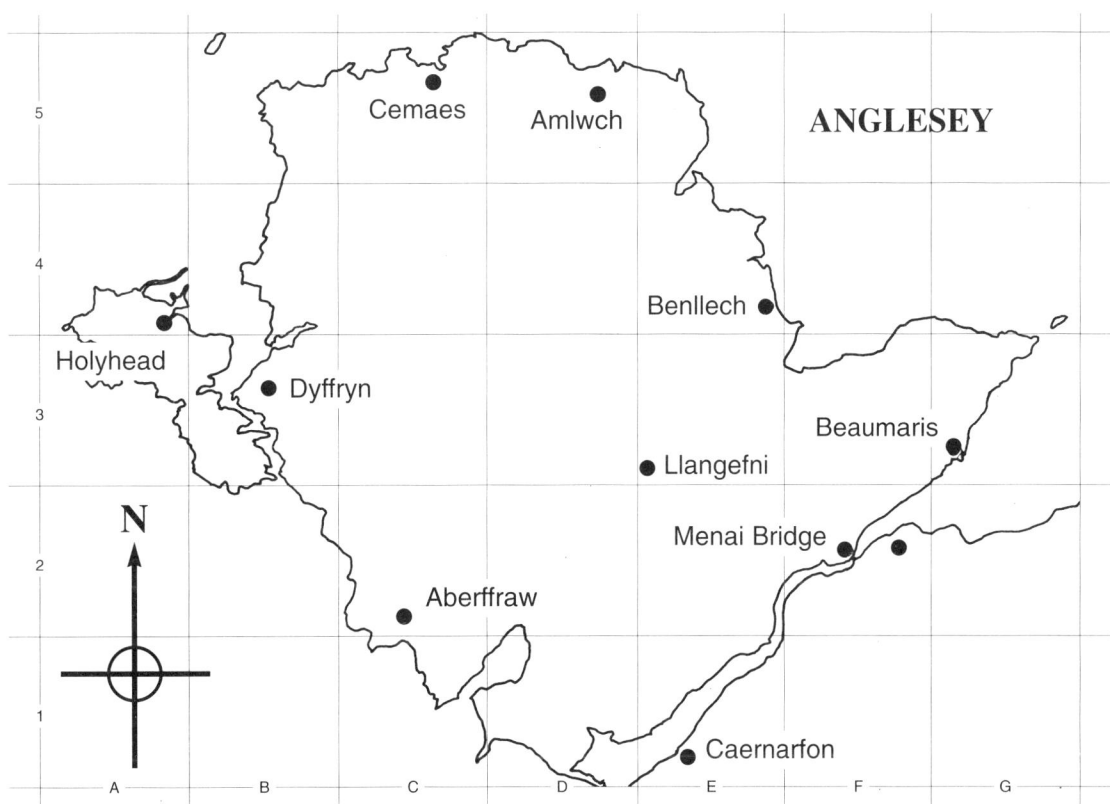

1. Write **G1** and **D4** in the squares with that reference.

2. Which town is in square **A4** ?

3. Is **Beaumaris** in square **F3** ?

4. What reference will these places have ?

    ➤ **Menai Bridge**   ➤ **Llangefni**

    ➤ **Amlwch**   ➤ **Benllech**

5. Mark *South, East and West* on the compass in squares **A1** & **A2**.

6. What is the most southerly town on the map ?

7. The city on the Welsh mainland opposite Menai Bridge is not named. What is it ?
   (Use an atlas, if necessary)

Maps (2)

All maps, atlases and town plans use a grid system to show where places are. In the index, the horizontal reference is often given first, followed by the vertical reference.

The most accurate atlases and maps give an exact reference to each place. This is done by numbering the **lines** in the grid, not the squares. The distance between the lines is divided into tenths *(like a metric ruler)*. The divisions are not usually marked on the grid and the reference does not include a decimal point.

e.g. If a place has the map reference **1538**, the first two numbers (**15**) refer to the horizontal scale and the last two numbers (**38**) refer to the vertical scale:

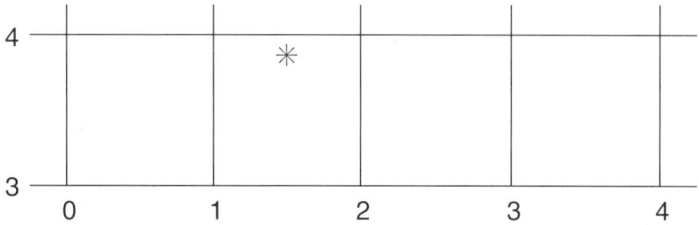

The map below shows the Isle of Man.

The mountain *Snaefell*, in the centre of the island, has the reference **4838**.

4838 = **4.8** along the horizontal *(bottom)* scale

3.8 up the vertical *(left-hand)* scale

1. Which place has the reference **5646** ?

2. Which place has the reference **5861** ?

3. Give the references for these places:
   a. Peel
   b. Port Erin
   c. Castletown
   d. Douglas

4. What reference could be given to the island off the south west coast of the Isle of Man ?

   (Use an atlas to find the name of the island)

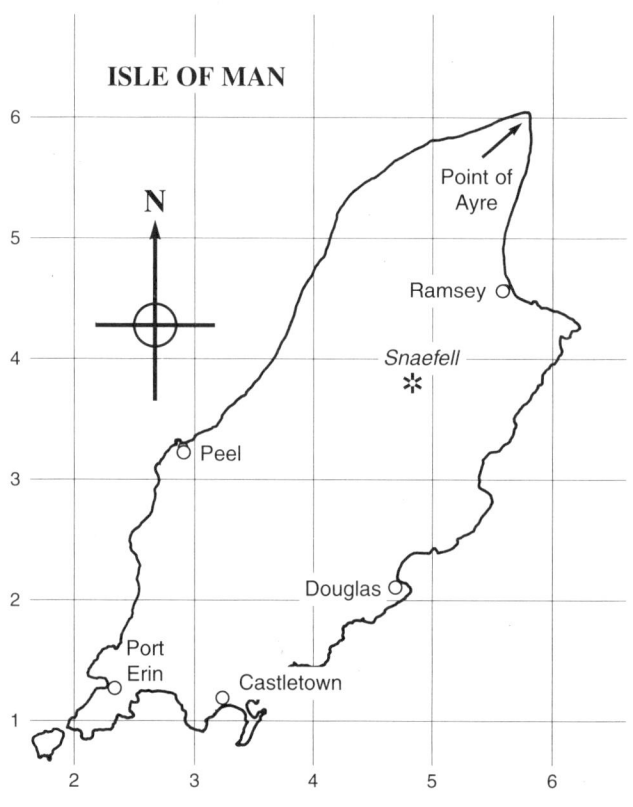

Maps (3)

A. *The map shows the main roads on the Isle of Wight off the south coast of England.*
The distances in miles between the points marked o are given.
   (**Note:** *The positions of* **Sandown** *and* **Shanklin** *have been slightly altered for the sake of simplicity.*)

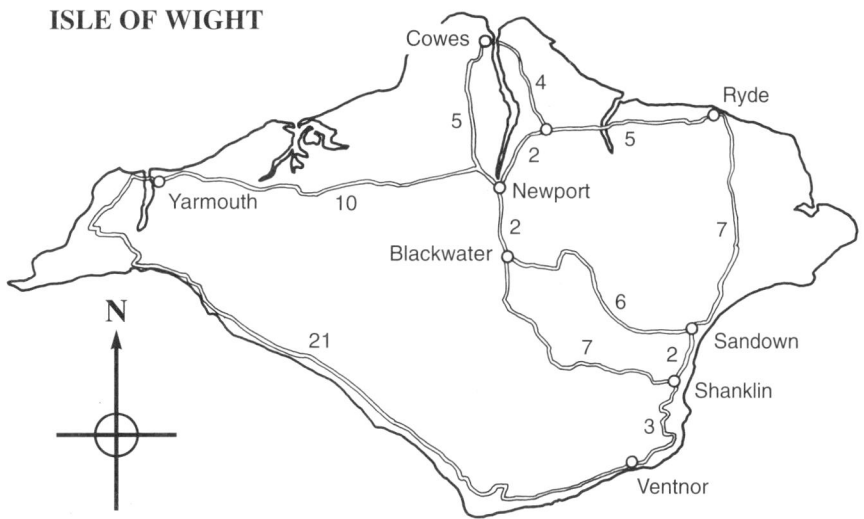

What is the distance between these places by the shortest route ?

1. Ryde and Newport   **2.** Ventnor and Ryde   **3.** Ventnor and Cowes

4. Yarmouth and Ryde   **5.** Ryde to Newport, Sandown and back to Ryde

6. Yarmouth (*via Newport*) to Ryde, Ventnor and back to Yarmouth

B. *The map shows some of the roads on the Isles of Lewis and Harris in the Outer Hebrides, North West Scotland.*

1. *From the scale shown on the map, estimate these distances:*

   a. Port Mholair to Stornoway

   b. Barvas to Carloway

   c. Aird Uig to Garynahine

2. *Give directions, spoken or written, as follows:*

   a. From Stornoway to Port of Ness

   b. From Carloway to Tarbert

   c. From Rodel to Hushinish

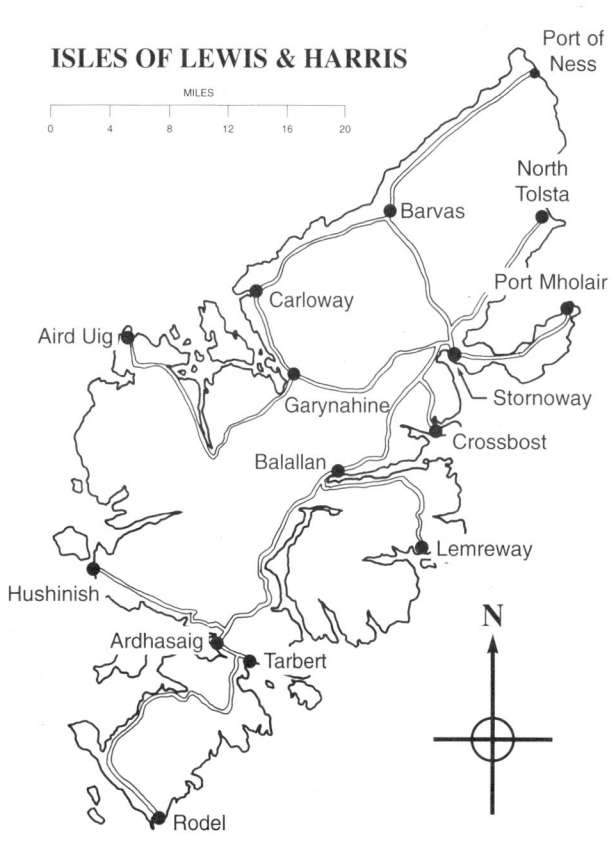

Directions and map references

A. Giving directions

Give directions from where you live to 3 of the places listed below. Include details of the distance and the length of time it would take to get to each place.

1. The nearest post box
2. The nearest telephone box
3. The nearest bus stop
4. The nearest petrol station
5. The nearest pub
6. The nearest shop
7. The nearest place of worship

B. Using a street plan

Using a street plan of your nearest large town or city:

1. Show how you would get from the main library to the police station.
2. Show how you would get from the bus station to the main post office.

C. Using a road atlas of Great Britain

Find the following places and write down the page number and map reference for each:

Chipping Ongar   Narberth   Dornoch

D. Using a road atlas for Ireland

Find the following places and write down the page number and map reference for each:

Magherafelt *(Northern Ireland)*   Tipperary *(Republic of Ireland)*

E. Using a world atlas

1. *Find and write down the country, page and map reference for these towns:*

Alleppey	Orlando	Murmansk
Kumasi	Rio Grande	Launceston

2. *Find these capital cities. Write down the map reference and the name of the country of which each is the capital.*

Copenhagen	Port of Spain	Dhaka (or *Dacca*)	Suva
Accra	Montevideo	Reykjavik	Kuala Lumpur

Shopping in metric

Since December 31st 1999, all food has been sold in metric weights and prices.

1 lb = 454 grams	2.2 lbs = 1 kilogram

A. What to ask for

As a rough guide to buying foods loose in metric, fill in the amounts in these statements from the list given in the box below.

50g	100g	250g	500g	750g	1kg	2kg

1. If you want a quarter of a pound, ask for _____ .

2. If you want half a pound, ask for _____ .

3. If you want a pound, ask for _____ .

4. If you want 2 pounds, ask for _____ .

Are the metric amounts slightly less or slightly more than the old pounds and ounces ?

B.

1. *Using a calculator, work out the price in £ per kg.*

   Price per lb: £0.15   0.25   0.50   1.00   1.50   2.00   3.00   4.00   5.00

   Price per kg:

2. *Draw a graph for the above prices, with £ per lb along the bottom (horizontal scale) and £ per kg up the side (vertical scale).*

3. *a.* *Fill in typical prices for these common foods which are often sold loose.*

   b. *Use the graph you have drawn to help you work out what the alternative prices will be.*

Potatoes	_____ per lb	_____ per kg
Cottage cheese	_____ per lb	_____ per kg
Sausage	_____ per lb	_____ per kg
Cooking apples	_____ per lb	_____ per kg
Onions	_____ per lb	_____ per kg
Pears	_____ per lb	_____ per kg
Bacon	_____ per lb	_____ per kg

Paying the Newsagent

A. You want to buy a few magazines to take to your aunt in hospital. These are the ones you think she might like:

Bella (64p)

Woman (£1.25)

Best (99p)

Woman's Weekly (60p)

Take a Puzzle (£1.60)

Good Housekeeping (£2.60)

1. What will be the cost of *Bella*, *Woman* and *Best* ?
   How much change will you get from £5 ?

2. Which 3 magazines will be the most expensive ?
   How much will the total be ?
   How much change will you get from £10 ?

3. Which 3 magazines will be the cheapest ?
   How much will the total be ?
   How much change will you get from £3 ?

B. You get the *Daily Mirror* (*Daily Record* in Scotland) at 32p every day for a week.

1. How much will the weekly paper bill be ?

2. How much change will you get from a £20 note ?

C. The monthly fishing magazine, *Trout and Salmon*, costs £2.60.

1. How much will it cost to buy it monthly for 6 months ?

2. A year's subscription, paid in advance, costs £31.20 including postage to your home. What will a subscription save you compared with buying it each month ?

D. Your weekly paper bill comes to £4.07. When you give a £10 note, you are asked if have the 7p. If you have it, what change will you then get ?

E. Most newsagents sell far more than just newspapers and magazines.

1. Make a list of 10 items which are commonly sold in a newsagent's, with their prices.

2. Add the 10 prices together.

Which kettle ?

These are five of the automatic cordless jug kettles advertised in a catalogue:

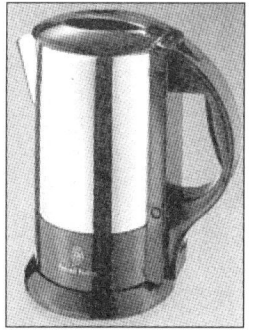 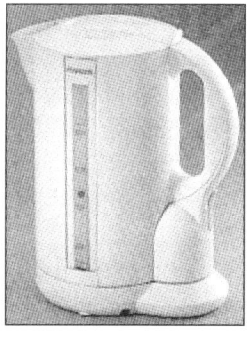   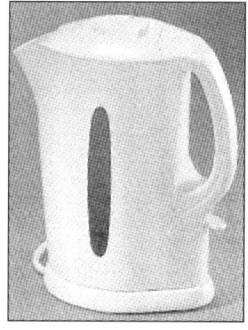

Russell Hobbs

Stainless steel cordless
jug kettle
Powerful concealed
element
1.7 litre capacity
360˚ cordless base

£29·50

KENWOOD

1.7 litre capacity
Wide-angled spout
Multi-position cord exit
Water level gauge
Stainless steel scale-
resistant element
Ergonomic switch design

£18·25

Rowenta

Only kettle that never
needs de-scaling!
Large easy-fill spout
1.5 litre capacity
2 water level gauges
Removable filter
Concealed element

£34·99

TEFAL

Twin water level gauges
3kW rapid boil concealed
element
360˚ cordless base
1.7 litre capacity
Extra fine mesh
removable filter
Hinged locking lid
Pilot light
Cord storage in base

£33·00

morphy richards

1.5 litre cordless kettle
Stylish water gauge
Removable lid for easy
cleaning
Cord storage in base
2 year guarantee

£16·75

A.
  1. Which is the cheapest kettle ?

  2. Which is the most expensive kettle ?

  3. Do all the kettles hold the same amount of water ?

  4. What are the possible drawbacks of the cheapest kettle ?

  5. Which kettle is likely to be the best quality ?

  6. Which kettle would you choose as the best buy ? Say why.

B. Compare the prices of these kettles with those available in other catalogues and in local shops.

C. Further exercises

Using a shopping catalogue, look up 3 of the following items:

| **Digital alarm clock** | **Baby buggy** | **Hair dryer** |
| **Exercise machine** | **Rechargeable shaver** | **Toaster** |

For each item, write down which is:

1. the cheapest model

2. the most expensive model

3. the one you think is the best buy

Which ear defender ?

A firm wants to buy some new ear muffs *(ear defenders)* for its workers. An industrial equipment catalogue gives 8 different models:

CLASSIC RANGE

MONZA
Top of range ear defender: high attenuation plus total comfort.
- Wire head band and liquid filled pads for comfort.
- Large, high impact polystyrene acoustic foam filled cups.

MONACO
High attenuation ear defender.
- Large, high impact polystyrene acoustic foam filled cups and foam pads.
- Wire head band.

SILVERSTONE
High attenuation ear defender.
- Large, high impact polystyrene acoustic foam filled cups.
- Padded plastic headband and liquid filled pads for comfort.

LE MANS
- High impact polystyrene acoustic foam filled cups and foam pads.
- Wire head band.

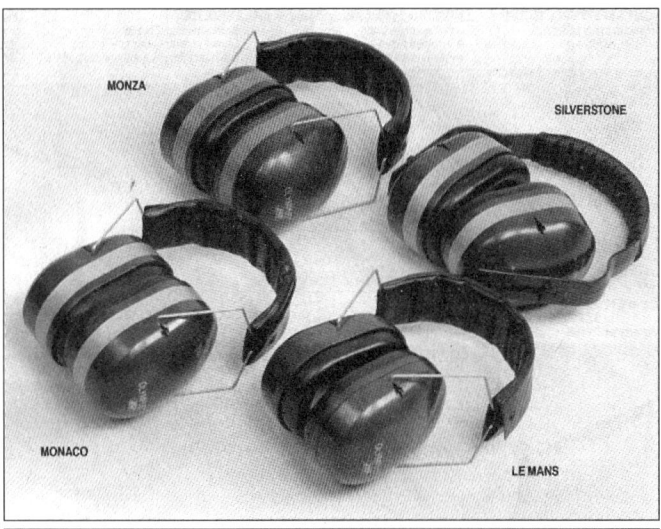

Ref.	Description	Price
101J060N	Monza	£14.45
102J060N	Monaco	£11.75
103J060N	Silverstone	£11.50
104J060N	Le Mans	£10.95

ECONOMY RANGE
- All ear defenders have impact polystyrene acoustic foam filled cups with foam pads.
- All models supplied in packs of 5.

GOODWOOD
High attenuation ear defender.
- Padded plastic headband.

BROOKLANDS
Mid-range ear defender.
- Padded plastic headband.

J-MUFF
- Ear cups fitted to glass re-inforced nylon headband, to minimise distortion.
- Colours: blue, white, orange or yellow. Please specify.
- Supplied in packs of 5 of a single colour.

E-MUFF
- Black ear cups with yellow foam pads fitted to glass filled nylon headband.

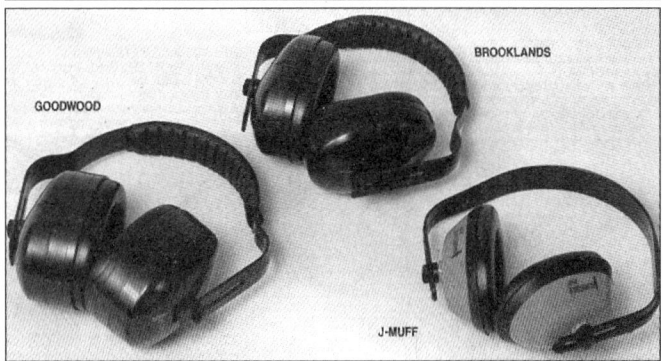

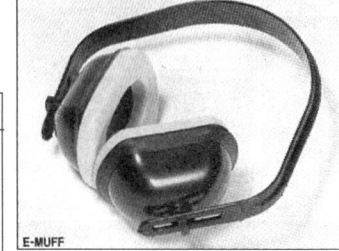

Ref.	Description	Price
105J060N	Goodwood	£28.00
106J060N	Brooklands	£24.70
100J060N	J-Muff	£14.05
172J060N	E-Muff	£13.40

1. How much is the most expensive ear defender ?

2 How much is the cheapest ear defender ? Give the price of a pack and of 1 ear defender.

3. How much does a **Goodwood** cost per ear defender ?

4. The firm decides to order 3 **Silverstone** ear defenders and 20 **J-Muff** ear defenders. Fill in the order form and prices *(without using a calculator)*.

Ref.	Description	Price per pack	No. of Packs	Price
			TOTAL *(ex-VAT)*	

Special Offers

For each of the Special Offers on chocolate biscuits, use a calculator to work out these prices:

A. *1 biscuit at normal price.*   **B.** *1 biscuit at the Special Offer price.*   **C.** *The saving per biscuit.*

Example

Special Offer: **4** *bars for the price of* **3.** **Pack: 40p**

A. 40p ÷ 3 = 13.3p   **B.** 40p ÷ 4 = 10.0p   **C.** 13.3p - 10.0p = 3.3p per biscuit

---

PENGUIN
1 free biscuit
8 biscuits for the normal price of 7
Pack price: **73p**

Blue Riband   *1 free biscuit*   *6 biscuits for the normal price of 5*   *Pack price:* **65p**

Tunnock's **CARAMEL** WAFER BISCUIT

1 free biscuit   **5** *biscuits for the normal price of* **4**
Normal pack price: **54p**

Time Out   *1 free biscuit*   **7** *biscuits for the normal price of 6*   *Pack price:* **89p**

Kit Kat
12 two-finger biscuits for 99p
Normal price: **16** *biscuits for* **£1.56p**

Wagon Wheels   *2 free biscuits*
8 biscuits for the normal price of 6   *Pack price:* **67p**

Rocky   *10 biscuits for £1.25*   *Normal price:* **6** *biscuits for 87p*

Chocolate Digestive Biscuits

25% free   *500g packet for the price of* **400g**
Packet price: **£1.20** *(approx. 24 biscuits per 400g)*

Maths and babies

There is a lot of maths involved in looking after babies and very young children. As soon as babies are born, they are measured and regular records should be kept of their growth.

The 4 main items which should be recorded are:

Head circumference   **Age**   **Weight**   **Length**

A. What is the head circumference ?

1. The height of the head from under the chin to the top of the head.

2. The width of the head from ear to ear.

3. The distance round the head including the ears.

4. The distance round the head at forehead level, like a hat measurement.

B. The figures below are for an *average* boy and girl.

Age (weeks):	0	4	8	12	16	20	24	28	32	36	40	44
Boy's weight (kg)	3.5	4.4	5.3	6.0	6.7	7.3	7.8	8.2	8.6	8.9	9.2	9.5
Girl's weight (kg)	3.3	4.2	5.1	5.8	6.4	7.0	7.4	7.8	8.2	8.6	8.8	9.2
Boy's length (cm)	51.2	54.8	57.9	60.7	63.2	65.2	67.0	68.5	69.9	71.2	72.3	73.6
Girl's length (cm)	50.3	53.6	56.6	59.3	61.6	63.6	65.2	66.7	68.0	69.3	70.6	71.8

*Using the above table of **average figures**, try to decide whether each of these statements is **True** or **False**:*

1. Boys are usually a little heavier than girls at birth.

2. Boys put on a lot more weight than girls in the first year.

3. Babies double their weight by age 16 weeks.

4. A 3.2 kg baby is about the same as a 7 lb baby.

5. Girls grow in length as fast as boys.

6. Babies double in length in a year.

7. Some girls are taller than boys at birth.

8. All babies should be over 50 cm in length at birth.

C. Further exercises

1. Plot the figures for either boys' or girls' weight on a graph. Estimate what the weight will be after a year.

2. Plot the figures for either boys' or girls' length on a graph. Estimate what the length will be after a year.

3. Make a list of some of the other maths measurements or calculations which parents may need to do during pregnancy and in the first 12 months of a baby's life.

Measuring for D.I.Y.

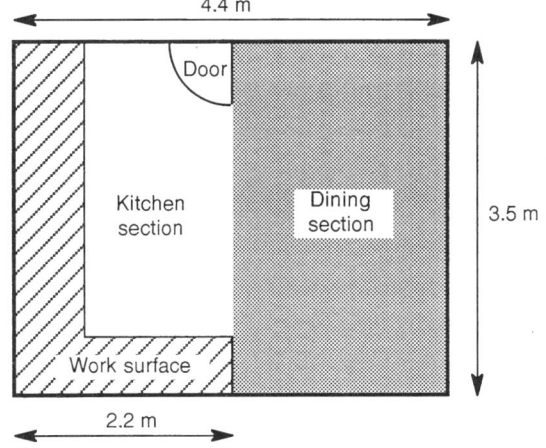

A kitchen/dining room measures 4.4 metres by 3.5 metres.

Half the area is the kitchen and the other half is a dining section with a table and chairs.

A. **1.** What is the floor area of the dining section ?

2. *The carpet for the dining section comes in a 4 metre width @ £4.99 per square metre.* What length from the carpet roll will need to be ordered ?

3. What will it cost to carpet the dining section ?

4. *Pine tables are available in the following sizes:*

750 x 750 mm	900 x 1200 mm
750 x 1200 mm	900 x 1500 mm
750 x 1500 mm	900 x 2100 mm

What size of table might be best for the dining section for a family of four ?
(Allow room for chairs and for getting round the table!)

B. *Most kitchen work surfaces come in a standard 600 mm width and in lengths of either 2 metres or 3 metres.*

1. What is the total length of work surface needed in the kitchen above ?

2. What is the area of the work surface as shown in the drawing ?

3. Work out which of the standard lengths of work surface will be needed ?

4. If the price of the work surfaces is £23.95 *(2m length)* and £30.50 *(3m length)*, what will the total cost be ?

C. *A notice board is needed for the kitchen. There is a choice of 2 types:*

- A ready-made cork board in a frame: *Size:* 600 x 800 mm   *Price:* **£7.49**
- Cork self-adhesive tiles which can be stuck on the wall:
  Size: 300 x 300 mm   *Price:* **£7.99** for a pack of 6.

1. Work out the area that the board would provide.

2. Work out the area that the tiles would provide.

3. Which is the better value for money ?

4. What are the advantages and disadvantages of each ?

Working out Benefits

A. Child Support Maintenance

A divorced mother has three children aged 8, 12 and 16. A sample calculation of the basic maintenance requirement for her children is shown below. Some of the figures are given *(December 2000 rates)*.

Fill in the blanks and work out the totals.

Personal allowance for each child under 16 (**£30.95**)   *(Fill in)*→

Personal allowance for each child aged 16 - 18 (**£31.75**)   *(Fill in)*→

Family expenses *(fixed sum)*   £14.25

Extra for single parent with child under 11   £52.20

                                             Total

Less child benefit (1st child, **£15.00**; others, **£10.00**)   *(Fill in)*→

                Weekly basic maintenance requirement

B. Working Families Tax Credit (WFTC)

WFTC is a tax-free benefit for people who work 16 hours or more and have children. The amounts of credit available in December 2000 are shown in the box below.

Basic tax credit *(one per family)*	**£53.15**
Child under 16	**£25.60**
Child aged 16 - 18 *(still in full-time education)*	**£26.35**
Extra credit for person working 30 hours or more per week	**£11.25**

Childcare credit - 70% of eligible childcare costs
(Eligible childcare costs up to maximum £100 for 1 child, or £150 for 2 or more children.)

Families with income *(including savings)* of £90 or less per week get the full WFTC.

Families with income *(including savings)* of more than £91.45 per week will have 55p deducted from WFTC for every £1 of income over £91.45.

Work out how much WFTC this family could get:

2 parents with 3 school children aged 10, 14 and 17. One parent works 21 hours per week and the family's total income is £90. There are no childcare costs and they have no savings.

C. Further exercises

Work out the weekly basic maintenance requirement *(as in Section A)* and the WFTC *(as in Section B)* for this family:

A single father bringing up 1 child, aged 6. He works 36 hours per week and earns £200 after tax He pays £60 per week for childcare and has no savings.

Reading timetables

A. *This is a train timetable for **The Cornishman**, which runs between Edinburgh and Penzance.*

Edinburgh	**0810**
Berwick-upon-Tweed	**0849**
Newcastle	**0937**
Durham	**0950**
Darlington	**1007**
York	**1039**
Leeds	**1111**
Wakefield	**1127**
Sheffield	**1159**
Derby	**1239**
Birmingham New St	**1319**
Bristol Temple Meads	**1452**
Exeter St. Davids	**1553**
Plymouth	**1649**
Liskeard	**1717**
Bodmin Parkway	**1730**
Par	**1741**
St. Austell	**1749**
Truro	**1807**
Redruth	**1819**
Camborne	**1826**
St. Erth	**1837**
Penzance	**1850**

1. Write down in words the times that the train leaves Edinburgh and arrives in Penzance.

2. What time *(on the 12-hour clock)* does the train arrive:
   a. In Birmingham ?
   b. In Bristol ?
   c. In Plymouth ?
   d. In Bodmin ?
   e. In Camborne ?

3. How long does it take to travel from:
   a. Edinburgh to Newcastle ?
   b. Leeds to Wakefield ?
   c. Birmingham to Truro ?
   d. Edinburgh to Penzance ?

B. *This is a timetable for National Express coaches between Torquay (Coach Station, Lymington) and London (Victoria Coach Station).*

Torquay Depart	London Arrive	London Depart	Torquay Arrive
0030	0620	0745	1320
0550	1120	1030	1555
0720	1310	1230	1735
0945	1515	1430	1945
1145	1710	1630	2125
1400	1950	1830	2325
1700	2220	2330	0510

1. If you want to get to London by 8 o'clock in the evening, which coach would you catch ?

2. If you want to return to Torquay by 3 o'clock in the afternoon, which coach would you catch ?

3. If you have a meeting in London, near Victoria, from 2 p.m. to 4 p.m., which coaches will get you from Torquay to London and back in a day ?

4. How long does the overnight coach take to get to London ?

5. Which is the quickest coach journey to London from Torquay ?

6. Which are the quickest coach journeys from London to Torquay ?

TV ratings

A. The table below shows the estimated viewing figures for four TV programmes during the autumn of 2000. The figures represent the number of viewers in millions.

Week ending:	1/10	8/10	15/10	22/10	29/10	5/11
Coronation Street	14.8	15.7	16.8	16.3	16.3	16.7
EastEnders	16.7	15.4	15.6	17.1	19.3	20.2
Emmerdale	10.7	10.5	11.2	10.7	10.8	11.3
Who Wants to Be a Millionaire ?	12.6	15.4	13.7	11.6	12.6	13.1

Answer these questions

1. All the programmes had more viewers for the week ending 5th November than they did for 1st October. Why might this be ?

2. Which is the most popular of the four TV programmes during the week ending *5/11* ?

3. In which week did the viewing figures for *Who Wants to Be a Millionaire?* go down by 2.1 million on the previous week ?

4. How accurate do you think the figures are ?
   Within: 5 million   2 million   1 million
   0.5 million   0.2 million   0.1 million

B. The top 10 programmes for the week ending 17th September were *(in alphabetical order)* as follows. The viewing figures *(in millions)* are in brackets.

   BBC News (Weds.) (8.6);   Big Brother (9.5);   Casualty (9.9);   Coronation Street (15.9);

   EastEnders (14.4);   Emmerdale (10.1);   My Fragile Heart (9.8);

   Neighbours (8.0);   The Bill (8.7);   Who Wants to Be a Millionaire? (13.1)

Put the programmes and their viewing figures into a table in order of popularity, then answer the questions.

1. Which programmes have around 10 million viewers ?

2. What is the difference in viewing figures *(in thousands)* between the bottom 2 programmes ?

3. How many of the programmes are 'soaps' ?

4. Why is the programme *Big Brother* so named ?

5. Why do *Neighbours, EastEnders* and *Coronation Street* have an advantage with their weekly viewing figures ?

C. Look up the most recent viewing figures, in *Radio Times* or some other TV listing magazines, and compare them with the ones above.

Football league tables

League tables are most common in sport, especially in team games; but they are now also used in other areas of life to give information, for example, on the performance of schools, hospitals and public services.

Sporting tables contain a mass of information hidden away in the small print. Below is the football league table for the F.A. Carling Premiership in England and Wales in December 2000.

A. Answer these questions on the league table.

1. Explain what these stand for:

    P   W   D   L   F   A   Pts

2. How many points are given:

    a. for a win ?   b. for a draw ?

3. Which team has:

    a. Won the most games ?

    b. Drawn the most games ?

    c. Lost the most games ?

    d. Scored the most goals ?

    e. Scored the fewest goals ?

    f. Conceded the most goals ?

    g. Conceded the fewest goals ?

    h. Played fewer games ?

	P	W	D	L	F	A	Pts	Goal Diff.
Manchester United	16	12	3	1	41	10	39	
Arsenal	16	9	4	3	24	13	34	
Leicester City	16	8	5	3	17	12	29	
Liverpool	16	8	3	5	32	23	27	
Ipswich Town	16	8	3	5	23	17	27	
Sunderland	16	7	5	4	17	16	26	
West Ham United	16	6	6	4	22	17	24	
Aston Villa	15	6	6	3	18	13	24	
Newcastle United	16	7	3	6	18	16	24	
Tottenham Hotspur	16	7	2	7	22	23	23	
Leeds United	15	6	4	5	22	22	22	
Charlton Athletic	16	6	3	7	21	24	21	
Everton	16	6	3	7	19	23	21	
Chelsea	16	5	5	6	28	23	20	
Southampton	16	4	5	7	21	28	17	
Manchester City	16	4	2	10	18	30	14	
Derby County	16	2	7	7	19	31	13	
Coventry City	16	3	3	10	15	32	12	
Middlesbrough	16	2	5	9	19	27	11	
Bradford City	16	2	5	9	9	25	11	

4. *If two clubs have the same number of points, the 'goal difference' decides which club is above the other. The right hand column in the table is headed **Goal Diff.***

    a. Subtract the goals **A** from the goals **F** for each team and write the answer in the **Goal Diff.** column. The top teams will have a + figure; the bottom teams will have a - figure.

    b. Which team has a zero goal difference ?

B. The current league tables

1. How many of the top six clubs in the above table are in the current top six ?

2. *Many league tables published in the newspapers separate the home and away games.*

    e.g. Tottenham Hotspur's record at the time of the above table was:

    Home: **W:** 7   **D:** 1   **L:** 0   **F:** 17   **A:** 6   *Away:* **W:** 0   **D:** 1   **L:** 7   **F:** 5   **A:** 17

    What does this tell you about Tottenham's record at home and away ?

3. a. Use the latest soccer league table (English and Welsh, Scottish or Irish, as appropriate) to answer questions *3. a to g* in *Section A*.

    b. Which team has the best record ?   i. Away   ii. Home

Calculating VAT

Businesses regularly have to calculate VAT on sales and purchases. VAT is currently (2000/01) charged at 17.5%. It is added to the cost of most items except food, books, newspapers, children's clothing and a few others.

A.

> **To calculate the VAT on an item where the price is quoted excluding VAT:**
> Using a calculator, *multiply* the price (ex VAT) by 17.5, then press the % key.
> *e.g.* What is the VAT on a batch of photocopy paper @ £208 *ex VAT* ?
> *Answer:* £208 x 17.5% = £36.40

Prices which exclude VAT

Using a calculator, work out the VAT on these items and services. The prices are quoted excluding VAT.

1.	Wall clock	£8.25	4.	Washing machine repair	£65
2.	Bicycle	£119	5.	3-piece suite	£808
3.	Toaster	£14.95	6.	Second-hand van	£2500

B.

> **To calculate the total cost including VAT when the price is quoted excluding VAT:**
> Using a calculator, *multiply* the price (ex VAT) by 1.175.
> *e.g.* What is the total cost of 30 reams of photocopy paper @ £208 *ex VAT* ?
> *Answer:* £208 x 1.175 = £244.40

Prices which exclude VAT

Using a calculator, work out the total cost, including VAT, of these items and services. The prices are quoted excluding VAT.

1.	Child's trike	£16.38	4.	Exercise machine	£179.38
2.	Microwave oven	£126.81	5.	Dental bill	£54.50
3.	Training course fee	£360	6.	Building contract	£169,000

C.

> **To calculate the basic cost excluding VAT when the price is quoted including VAT:**
> Using a calculator, *divide* the cost (inc. VAT) by 1.175
> *e.g.* What is the cost excluding VAT of a batch of photocopy paper @ £244.40 *inc. VAT* ?
> *Answer:* £244.40 ÷ 1.175 = £208.00

Prices which include VAT

Using a calculator, work out the cost excluding VAT, and the VAT itself, on these items and services. The prices are quoted including VAT.

1.	Stationery items	£7.29	4.	Conservatory	£4,218.50
2.	Hi-Fi system	£349.95	5.	Pencil	8p
3.	Computer	£899	6.	Volvo Truck	£77,245.67

Answers

p. 6 Adding and Subtracting - Figures
A. 189, 216, 127, 255, 131 **Total:** 918   **B. 1.** £274.92   **2.** £63.80   **3.** £211.12

p. 7 Dates and Calendars
A. **1.** 7   **2.** 30 *or* 31, *except* February, 28 (*Leap Year*, 29)   **3.** 52   **4.** 12   **5.** 365 (*Leap Year*, 366)   **6.** Every 4 years

B. **1. a.** April   **b.** In the U.S.A. 4/9/05 = April 9th 2005 (*9/4/05 would be September 4th*)

    2. March, November, May, September, December, January, July, February, October, June, August

C. **1.** Wednesday   **2.** 5   **3.** August 4th   **4.** August 29th   **5.** 5   **6.** Monday   **7.** Sunday   **8.** 3

    9. a. Bank Holiday in England, Wales and Northern Ireland only

      b. In Scotland & the Republic of Ireland, Bank Holiday is the 1st Monday in August.

p. 8 Time and Clocks (1)
A. **1.** 60   **2.** 24   **3.** 60   **4.** 12, a.m. & 12, p.m.   **5.** Twice (*back 1 hr in late October; forward 1 hr in late March*)

B. **1.** 10.30 a.m., 1030; 1.50 a.m., 0150; 7.15 p.m., 1915; 9.45 p.m., 2145; 12.00 noon, 1200;
    11.02 a.m., 1102; 1.55 p.m., 1355; 3.40 p.m., 1540; 12.00 midnight, 2400 *or* 0000; 5.43 a.m., 0543

    2. 7.30 a.m.; 11.22 a.m.; 11.06 p.m.; 6.53 p.m.; 4.40 p.m.; 12.00 noon; 12.01 a.m.; 9.15 p.m.

p. 9 Time and Clocks (2)
A. **1.** 6.15 p.m. - 8.00 p.m.   1815 - 2000   (In practice, allow 15 mins. extra to cover any delays!)

    2. 11.35 a.m. - 12.45 p.m.   1135 - 1245   (In practice, allow 15 mins. extra to cover any delays!)

    3. Allow about 2 hrs from start to finish (including about 15 mins for water to come to boil).

    4. 6.45 - 8.45 a.m.; 5.30 p.m. - 12.00 a.m.   0645 - 0845; 1730 - 0000

    5. Exact times are: 11.50 p.m. - 1.50 a.m.   2350 - 0150
    but settings might be: 11.45 p.m. - 2.10 a.m.   2345 - 0210   to be on the safe side.

    6. 5.50 p.m. - 1.00 a.m.   1750 - 0100

p. 10 Adding and Subtracting - Hours
A. **1.** 15 mins; 0; 1 hr 15 mins; 25mins; 2 hrs 45 mins; 0   **2.** Total overtime: 4 hrs 40 mins

    3. Overtime paid: 5 hours   **B. 1.** 6 $^3/_4$; 4 $^1/_4$; 14 $^1/_4$; 15 $^3/_4$; 5   **2.** *Total:* 46 hrs

p. 12 Calculating the area
A. **1.** 407.6 m^2 **2.** 486 yd^2 **3.** A square metre   **B. 1.** 19 m^2 **2.** 76

C. **1.** 152.5 m^2 **2.** 4 tins (without allowance for doors and windows).

D. **1.** 0.96 m^2 **2.** 14 boards (*The bin will be 7 boards high; each board, cut in half, will do 2 sides*)

E. **1.** 1.5 m^2 **2.** 0.6 m^2 **3.** 252 bricks (although, in theory, 240 would do because bricks are 100 mm thick)

F. **1.** Yes - just about.   **2.** Not quite - the area covered will be roughly circular with a diameter of about 6
    metres, so the ends and corners will not be covered if the sensor is in the centre of the room.

p. 13 Calculating the volume
A. **1.** 11.5 m^3 **2.** 48 m^3 **3.** 21.6 m^3 **4.** 15.6 m^3 **5.** 25.9 m^3 **6.** 17.5 m^3 **7.** 11.5 m^3

B. **1.** 2.1m^3 **2.** 2145 litres   **C. 1.** 117 litres   **2.** 0.117 m^3 **D.** 1.4 m^3 if filled to the top.

p. 14 Fractions
B. (*clockwise from top left*) $^1/_2$; $^1/_2$; $^1/_4$; $^3/_4$; $^1/_8$; $^3/_8$; $^2/_5$; $^1/_3$; $^5/_8$; $^2/_3$; $^1/_3$

p. 15 Proportion
A. 1 in 2 marriages; 7 out of 10 apples; 1 in12 is a vegetarian; 1 in 7 small businesses close

B. 1 in 2 (10 in 20) is female; 1 in 5 (4 in 20) is wearing a hat.

C. **1.** 3 in 5 (6 in 10) point to the right   **2.** 2 in 5 (4 in 10) point to the left.

D. **1.** 1 in 10 points East   **2.** 3 in 10 point West   **3.** 1 in 5 (2 in 10) points South   **4.** 2 in 5 (4 in 10) point North

E. **1.** 2 in 5 (4 in 10) are **b**   **2.** 3 in 5 (6 in 10) are smiling

p. 16 Decimals
A. *l. to r.:* 0.5; 0.25; 0.33; 0.2; 0.1; 0.01   **B. 1.** £158.92   **2.** £196.38   **3.** £200.73   **4.** £146.13

C. *l. to r.:* 62.7; 3.17; 1890; £37.80; £52.00; £38.00; 74.8 m^2; 96,710 cm; 4,601g; 7,801,001; 54,962.02;
    £18, 364,279

Answers

p. 17 Percentages
A. 16%; 72%; 45%; 8%; 99%   **B. 1.** 3%; 12%; 87%; 50%
C. 1. Probably true in most cases.   **2. to 7.** All true

p. 18 Rounding numbers
A. 90, 40, 160, 30, 480, 10, 1970, 520, 70, 100, 100   **B.** 100, 600, 700, 300, 400, 900, 800, 200, 500, 1800
C. 9, 17, 11, 3, 95, 165, 100, 0, 8, 108   **D. 1.** 4, 14.6, 7.1, 271.6, 3.2, 8.5, 19.0, 19.3, 111.8

p. 19 Rounding numbers on a calculator
A. £1.65; £684.97; £8.00; £1324.64; £0.33; 5p; 96p
B. 1. 4 people get £416.67; 2 people get £416.66   **2.** Two pay £6.33; one pays £6.34   **3.** £55.50   **4.** £71.14
5. £13.00   **6.** 12p   **7. a.** £2.66   **b.** £7.35   **c.** Nothing   **8. a.** £8.95; £170.00   **b.** £129.49; £2460.28   **c.**
£18.21; £345.90   **9. a.** £9.99   **b.** £249.83

p. 20 Estimating (1)
Exact answer (where there is one): **1.** £2.28   **2.** Yes - 10kg would cost £3.17 at small bag price
3. 2.15 p.m. at the earliest, allowing a short wait for the bus.   **4.** £619.99   **5.** £177.60   **6.** 33.6 tonnes
7. No, it will have to go on the roof rack.   **8.** No - even a tall person wouldn't reach it safely.
9. £102.01   **10.** 3.7 - 4.3 metres *or* 12' - 14'

p. 21 Estimating (2)
(Answers to 1, 6, 7, 9 & 10 omitted on purpose) **2.** £8.36   **3.** 65p   **4.** They've got a chance, but it isn't
very likely.   **5.** In theory, £569; but often a 3rd week is much more expensive than a 2nd one.
8. a. 53p **b.** £1.50 (*Note: Actual bill may be: a. 57p b. £1.61, because some phone companies charge part of a
minute as a full minute.*)

p. 24 Scales and Dials
A. 2. 60g is the basic weight limit of a 1st or 2nd class U.K. Inland letter.

p. 25 Thermometer and Radio Scales
A. 1. 37.0°C 98.6F   **2.** 37.6°C 99.7°F   **3.** 38.6°C; 103.2°F   **4.** About 40°C or over (104°F or over)
B. 1. a. 96.5   **b.** 750   **c.** 194

p. 26 Bar Charts
B. 1. True   **2.** False   **3.** Maybe   **4.** Maybe   **5.** Maybe   **6.** Maybe (*All the statements from 3. to 6. could
account for the larger prison population in the U.K.*)
Bar codes are the product price code bands for laser scanners in shops *etc.*

p. 28 Line graphs
A. 1. 3p   **2.** 16p   **3.** 25p   **4.** 27p

p. 29 Maps (1)
2. Holyhead   **3.** No - **G3**   **4.** Menai Bridge: **F2**; Llangefni: **E3**; Amlwch: **D5**; Benllech: **E4**
6. Caernarfon   **7.** Bangor

p. 30 Maps (2)
1. Ramsey   **2.** Point of Ayre   **3. a.** Peel: 2933   **b.** Port Erin: 2313   **c.** Castletown: 3212
d. Douglas: 4721   **4.** 1708: Calf of Man

p. 31 Maps (3)
A. 1. 7 miles   **2.** 12 miles   **3.** 17 miles   **4.** 17 miles   **5.** 22 miles   **6.** 50 miles
B. 1. a. 11 miles   **b.** 10 miles   **c.** 21 miles (*actual mileages taken from Ordnance Survey atlas*)

p. 33 Shopping in metric
A. 1. 100g   **2.** 250g   **3.** 500g   **4.** 1kg   (**1.** is slightly less; **2.**, **3.**, **4.** are slightly more than the old lbs/oz)
B. 1. 33p/kg; 55p/kg; £1.10/kg; £2.20/kg; £3.30/kg; £4.40/kg; £6.60/kg; £8.80/kg; £11.00/kg

p. 34 Paying the Newsagent
A. 1. £2.88; £2.12 change   **2.** *Good Housekeeping, Take a Puzzle, Woman:* Cost - £5.45; Change - £4.55
3. *Bella, Best, Woman's Weekly:* Cost - £2.23; Change - 77p
B. 1. £1.92 (*not Sundays!*)   **2.** £18.08   **C. 1.** £15.60   **2.** Nothing - but postage free.   **D.** £6.00

Answers

p. 35 Which kettle ?

A. **1.** Morphy Richards **2.** Rowenta **3.** No - 2 hold 1.5 litres; 3 hold 1.7 litres **4.** Smaller capacity, *plus* features not listed (*e.g.* filter; extra water gauge; fast boil element; multi-position cord exit etc.)
5. Not necessarily the most expensive! (see independent tests in *Which?* magazine)

p. 36 Which ear defender ?

1. Monza: £14.45 **2.** E-Muff: £13.40 per pack of 5; £2.68 each **3.** £5.60 **4.** Total £90.70

p. 37 Special Offers

(from top) **Penguin:** 10.4p; 9.1p; 1.3p **Blue Riband:** 13.0p; 10.8p; 2.2p
Tunnock's Caramel Wafer: 13.5p; 10.8p; 2.7p **Time Out:** 14.8p; 12.7p; 2.1p **Kit Kat:** 9.8p; 8.3p; 1.5p
Wagon Wheels: 11.2p; 8.4p; 2.8p **Rocky:** 14.5p; 12.5p; 2.0p **Chocolate Digestives:** 5.0p; 4.0p; 1.0p

p. 38 Maths and babies

A. 4. **B.** **1.** True **2.** False **3.** Usually false **4.** True **5.** True **6.** False **7.** True **8.** False
C. **1.** Boy: 10 kg; Girl: 9.7 kg **2.** Boy: 75.8 cm; Girl: 74.0 cm

p. 39 Measuring for D.I.Y.

A. **1.** 7.7 m² **2.** 2.3 m minimum; 2.5 m, possibly **3.** 4m x 2.3m = £45.91; *or* 4 m x 2.5 m = £49.90
4. 750 x 1500 would probably be the maximum size, to allow some space round it.
B. **1.** 5.1 m **2.** 3.1 m² **3.** 2 x 3 m **4.** £61.00
C. **1.** 0.48 m² **2.** 0.54 m² **3.** The tiles are slightly cheaper per sq.metre, but the board is a cheaper buy.

p. 40 Working out Benefits

A. Weekly basic maintenance requirement: £125.10
B. WFTC: £130.70
C. Weekly basic maintenance requirement: £82.40; WFTC: £72.60

p. 41 Reading Timetables

A. **1.** The train leaves Edinburgh at ten minutes past eight in the morning and arrives at Penzance at ten minutes to seven in the evening.
2. a. 1.19 p.m. **b.** 2.52 p.m. **c.** 4.49 p.m. **d.** 5.30 p.m. **e.** 6.26 p.m.
3. a. 1 hr 27 mins **b.** 16 mins **c.** 4 hrs 48 mins **d.** 10 hrs 40 mins
B. **1.** 2.00 p.m. (1400) - but you might be late! **2.** 7.45 a.m. (0745) **3.** 7.20 a.m. (0720) up; 4.30 p.m. (1630) back *(with luck!)* **4.** 5 hrs 50 mins **5.** 5.00 p.m. (1700) takes 5 hrs 20 mins. **6.** 4.30 p.m. (1630) & 6.30 p.m. (1830) both take 4 hrs 55 mins.

p. 42 TV ratings

A. **1.** TV ratings are said to go up as the dark nights draw in. **2.** *EastEnders* **3.** *22nd October*
B. **1.** *Emmerdale* & *Casualty* **2.** Six hundred thousand **3.** Four (*Coronation Street; EastEnders; Emmerdale; Neighbours*). TV magazines usually describe *Casualty* and *The Bill* as 'Drama'. **4.** The title refers to the phrase *Big Brother is watching you* in George Orwell's novel, *1984*. It was used on posters as a warning that people could not escape the eye of the authorities. **5.** Because their ratings include repeat showings.

p. 43 Football league tables

A. **1.** Played; Won; Drawn; Lost; (Goals) For; (Goals) Against; Points
2. 3 for a win; 1 for a draw
3. a. Manchester United **b.** Derby County **c.** Coventry City & Manchester City **d.** Manchester United **e.** Bradford City **f.** Coventry City **g.** Manchester United **h.** Aston Villa & Leeds United
4. a. 31, 11, 5, 9, 6, 1, 5, 5, 2, -1, 0, -3, -4, 5, -7, -12, -12, -17, -8, -16 **b.** Leeds United
B. **2.** They're good at home, terrible away!

p. 44 Calculating VAT

A. **1.** £1.44 **2.** £20.83 **3.** £2.62 **4.** £11.38 **5.** £141.40 **6.** £437.50
B. **1.** £19.25 **2.** £149.00 **3.** £423.00 **4.** £210.77 **5.** £64.04 **6.** £198,575.00
C. **1.** £6.20 *inc.VAT* (VAT = £1.09) **2.** £297.83 (£52.12) **3.** £765.11 (£133.89) **4.** £3,590.21 (£628.29)
5. 7p (1p) **6.** £65,741.00 (£11,504.67)

Index